The Student Pilot's Flight Manual Syllabus

THE STUDENT PILOT'S FLIGHT MANUAL SYLLABUS

Second Edition

A Flight Instructor Checklist and
Student Pilot Guide
from the First Flight to the Private Certificate

William K. Kershner

IOWA STATE UNIVERSITY PRESS / AMES

Printed on acid-free paper in the United States of America
First edition, 1997
Second edition, 1998

International Standard Book Number: 0-8138-2927-5

Last digit is the print number: 9 8 7 6 5 4 3 2 1

CONTENTS

*If an approved flight simulator or other approved flight training device and authorized instructor are available, FARs 61.4 and 61.109 permit them to be used for these Units.

TRAINING RECORDS

Reference: FAR 141.101 (paraphrased)

Each holder of a pilot school or provisional pilot school certificate shall establish and maintain a current and accurate record of the participation and accomplishment of each student enrolled in an *approved* (141) course of training. (The student's logbook is not acceptable for this record.)

A copy of this syllabus and all required information should be kept for each student. This page should be filled out and the approved school authority, upon request of a student, should make a copy of this record available to the student.

Student's Name_____

Date of Enrollment_____

Date of Termination (Student Drop-Out)_____

Date of Transfer to Another School_____

Date and Grade of School STAGE 1 Presolo Knowledge Test_____

Date and Grade of School STAGE 2 Written Test_____

Date and Grade of School Postsolo Cross-Country Knowledge and Practical Test

_____

Date and Grade of School STAGE 3 Written Test_____

Date and Grade of School STAGE 4 Written Test_____

Date and Grade of FAA Knowledge Test _____

Date and Grade of School Final Practical Test_____

Date of Certification/Graduation and Result of FAA Practical Test

Date_____ Pass_____ Fail_____

Signature of Chief, or Other Authorized, Instructor

DATE_____SIGNED_____

Certificate No./Expiration_____

HOURLY BREAKDOWN: GROUND INSTRUCTION AND FLIGHTS

STAGE 1 - PRESOLO

Unit	Ground	Dual	Solo
1	2.0	1.5	-
2	1.0	1.5	-
3	1.0	1.0	-
4	2.0	1.0	-
5	1.0	1.0	-
6	1.0	1.0	-
7	1.0	1.0	-
8	1.0	2.0	-
9	1.0	1.0	-
10	1.0 Presolo Test	1.0	1.0
	12.0 Hours	12.0 Hours Minimum	1.0

STAGE 2 - IMMEDIATE POSTSOLO

Unit	Ground	Dual	Solo
1	0.5	1.0	1.0
2	-	-	1.0
3	1.0	1.0	1.0
4	1.0	-	-
5	0.5	-	1.5
6	1.0	1.0 (Hood work)	-
7	1.0	1.0	-
8	0.5	-	1.0
9	1.0	1.5	-
10	0.5	-	1.5
11	1.0	1.5 (Hood work)	-
12	1.0	-	1.0
13	1.0 - STAGE 2 Test	-	-
	10.0 Hours	7.0 Hours	8.0 Hours

STAGE 3 - CROSS-COUNTRY

Unit	Ground	Dual	Solo
1	2.0	-	2.0
2	3.0	-	-
3	1.5	1.0 (Night)	-
4	3.0	-	-
5	1.5	2.5 (XC)	-
6	1.0	-	3.0 (XC)
7		1.5	-
8	1.0	1.5 (Hood work)	-
9	1.0	-	3.0 (XC)
10	1.0	2.5 (Night)	-
11	1.0	-	1.0
12	1.0	2.0 (XC)	1.0
13	6.0	-	-
14	1.0	-	4.0 (XC)
15	1.0 - STAGE 3 Test	-	-
	25.0 Hours	11.0 Hours	14.0 Hours

STAGE 4 - PREPARATION FOR THE KNOWLEDGE TEST

Ground Instruction	9.0
School Knowledge Test	2.0
FAA Test	3.0
	14.0 Hours

STAGE 5 - PRACTICAL TEST PREPARATION

Unit	Ground	Dual	Solo
1	3.0	2.5	-
2	-	-	1.0
3	-	-	1.0
4	1.0	1.5	-
5	1.0	1.0	-
6	2.0	-	-
7 School Final Check	1.0	2.0 Final STAGE Check	-
8 FAA Check	2.0	2.0 FAA Check	-
	10.0 Hours	9.0 Hours	2.0 Hours

Total Hours	Ground	Dual	Solo
	71.0	39.0	25.0

BOOKS AND EQUIPMENT

This syllabus is based on use of the following books and equipment. They should be available for study and use by the student pilot and instructor during this course:

The Student Pilot's Flight Manual, Seventh Edition (SPFM)

The Flight Instructor's Manual, Third Edition (FIM) for use by the flight instructor to cross-reference with *SPFM.*

Pilot's Operating Handbook (POH)

Plotter

Flight computer (E6-B or electronic type)

Aeronautical Information Manual (AIM)

Federal Aviation Regulations: Parts 1, 61, 71, 91, and 141

Aviation Weather (AW) AC 00-6

Aviation Weather Services (AWS) AC 00-45

Current sectional chart(s) for the cross-country flights and the local area

Airport/Facility Directory (A/FD) for the airports in the training course

Pilot logbook

Practical Test Standards (latest) for Private Pilot, ASEL

The flight school or instructor should have available a library of Advisory Circulars that apply to student and private pilots.

INTRODUCTION

This syllabus is written to meet the requirements of FAR Parts 61, 141, and 142, and is intended to be used primarily as a reference, guide, and record for the flight instructor (for use with the student) from the first flight through the private pilot (ASEL) Knowledge and Practical Tests and Certification. The various elements are addressed for instructor action unless the student is specifically mentioned.

By design this is a summary, in order, of the steps the aspiring private pilot must take to successfully complete the course of ground and flight instruction and is to be used by the flight instructor as a checklist. For each subject reference is made to the more detailed information in *The Student Pilot's Flight Manual (SPFM)* or *The Flight Instructor's Manual (FIM)* so that both the student and instructor may read more details about a particular subject or flight maneuver. *However, this is intended to be more than a bare-bones syllabus with reminders, so memory joggers (cryptic comments) are included to help the flight instructor review what was learned from SPFM, FIM, and the other references cited here.* The instructor should fill out and keep a copy of this syllabus for each student as a back-up reference. Each student should have his or her own copy, as well, for possible transfer purposes.

This syllabus is broken down into Units rather than flights or hours because of the variations in student progress, which may be a function of schedule. For instance, bad weather, illness, or other factors may cause a large time gap between flights, requiring extra flight or groundschool review. The instructor using this syllabus will in some cases divide the Units into several sessions for easier handling; some suggested "break" points have been included (STAGE 3, Unit 4 is one example).

As the elements of each Unit are completed to the satisfaction of both the student and instructor, the next Unit should be introduced. In some cases the instructor may have the student go on to a following Unit in the same STAGE because of weather problems and, for instance, do pattern work rather than fly to the practice area. The incomplete Unit then should be completed as soon as possible.

Certain maneuvers are referred to in several Units and the user may think, "This was mentioned earlier." Yes, it was; but as a memory jogger and review, those items are repeated.

The time noted for each Unit, ground instruction or flight, is an estimate since one flight school may be at an uncontrolled airport and have its practice area very close, while another school may be located at a busy controlled airport, thus, much more time is needed to complete the Unit. Suggested ground instruction time for each Unit includes pre- and postflight briefings.

By checking off the items covered, the instructor and student each can review the syllabus to confirm the areas completed (or those that need to be reviewed). The student should initial each Unit to confirm its completion.

There will not be a repeat of "completion standards" with each Unit. The standards for the various Units will be complete when the student has performed the requirements for that Unit to the satisfaction of the flight or ground instructor, as applicable.

At the end of each Unit the Assigned Reading is in preparation for the next Unit.

The minimum requirements for a Part 141 school are 35 hours of ground instruction and 35 hours of flight training, with 20 hours of flight training from a CFI. The minimum requirements for Part 61 operations are 40 hours flight time, including at least 20 hours of flight instruction from a certified flight instructor and 10 hours of solo flight with the breakdowns as listed in FAR 61.109.

The average flight time for the certification of a private pilot (Airplane SEL) in the United States is approaching 75 hours, and this syllabus is intended to cover more than the bare minimums as cited by either Part 61 or 141. This syllabus requires a *minimum* of 12 hours of dual before solo (the average student will need more time than this); with the five STAGES included, the course is set up for a total of 71 hours of ground instruction or oral exams, 39 hours of dual flight (including check flights), and 25 hours of solo flight, for a combined total of 135 hours. These times can be modified to fit a particular airplane or local condition but still must meet the minimum number of hours of ground instruction, dual or solo. The goal of the program is not to just provided a minimum number of hours of instruction but to train a safe and competent pilot, and it is up to the instructor(s) to assure that flight times and more-than-minimum performance standards are met.

The cross-country requirements cited in STAGE 3 are a case in point. This writer believes that there is not enough solo cross-country, both in times and distances, required by the new (1997) FAR 61 and so has kept the "older" requirements in this area of instruction. The instructor/user, of course, may limit the experience to that currently required by FAR 61.

Note that up to 2.5 hours of instruction may be given in a flight simulator or flight training device representing an airplane from an authorized instructor and credited toward the total hours required by FAR 61.109. A maximum of 5 hours of instruction in a simulator or flight training device representing an airplane may be credited toward the total hours required by FAR 61.109 if the instruction is accomplished in a course conducted under FAR 142. The instruction flights that may use a flight simulator are so indicated in this syllabus.

Much thanks is owed to Paul A. Craig, author of several aviation textbooks and chief instructor of a major flight school, for his keen eye and ability to find gaps in the material required. His comments and suggestions were most valuable. I also want to thank Ellen Roberts, a flight instructor whose techniques and methods should be copied by more of us. James E. Perkins, Aviation Safety Inspector at the Nashville FSDO, was most helpful with his review of the manuscript and comments and suggestions. My son, Bill, who flies for American Airlines and is a former flight instructor, reviewed the manuscript and made valuable suggestions also. Thanks to ISU Press editor Lynne Bishop, whose humor and always good help is much appreciated. And, as with my other books, my wife Betty gave most invaluable typing and editing help.

I would appreciate comments and suggestions to make this syllabus more useful to the flight instructor *and* to the student.

William K. Kershner CFI 442723
P.O. Box 3266
Sewanee, TN 37375

THE STUDENT PILOT'S
FLIGHT MANUAL
SYLLABUS

STAGE 1

PRESOLO

This syllabus is more detailed at the beginning, gradually tapering off as the student approaches solo. In the last few flights before solo, the instructor acts more or less as a check pilot, with occasional comments or instruction, as needed. By the tenth flight, most of the necessary instruction has been given and only reviews and reminders are necessary as the student gains confidence and develops the skills for taking the airplane around the airport solo.

UNIT 1 - Introduction of Effects of Controls and the Four Fundamentals

■ **Ground Instruction. 2.0 Hours.** The instructor should discuss the following topics in the office or classroom before flying. The amount of detail before the first flight will depend on the student's background and time elements. The instructor may set aside more time for this first ground session and may not fly at this first meeting. Some points may be reserved for Units 2 or 3. The student should already have read *SPFM*, Chapters 2-9. The instructor's reference is *FIM*, Chapters 4 and 5.

____Effects of Controls (*SPFM*, Chapter 8).

    ____Elevators.

        ____Pitch control.

        ____Control angle of attack and airspeed.

        ____Use of elevator trim in various phases of flight. The importance of trimming to relieve pressure on the control. The trim tab is the "poor man's autopilot," and its primary purpose is to maintain certain attitudes/airspeeds without the pilot having to continually exert control pressure. Some more-advanced airplanes have trim available for rudder and ailerons.

        ____Change of elevator effectiveness with airspeed. Slow flight or flight at minimum airspeeds or maximum angles of attack (*SPFM*, Chapter 2, - "*Angle of Attack*") will be covered again later in the course.

    ____Ailerons.

        ____Roll control.

        ____Why and how an airplane turns; the horizontal lift vector.

        ____Adverse yaw; why rudder is used with ailerons in starting or completing turns.

    ____Rudder.

        ____Yaw control.

        ____Use in turns.

        ____Use in slips. Covered in more detail in *SPFM*, Chapter 13, "Slip".

____Correction for the left-turning forces in a climb or slow flight.

____Primary use for steering on the ground for some airplanes. Introduction of nosewheel or tailwheel steering in taxiing.

____Adverse yaw. A review, as given in the aileron discussion.

____Rudder tab or rudder trim use.

____Throttle.

____Power control.

____How it works: Move it forward (open) for more power, retard it (close) for less power.

____The climb and descent control.

____Flaps.

____Why they are used.

____Lift and drag effects.

____Stall speed effects.

____Approach angle changes with flaps.

____Four Fundamentals: Turns, Climbs, Straight and Level, and Glides (*SPFM*, Chapter 9).

____Turns.

____Entering the turn; application of ailerons and rudder.

____Why is back pressure required in the turn?

____Three aspects of the turn (ailerons, rudder, and elevator).

____Load factors in the turn.

____Rolling out of the turn; the need for relaxing back pressure. The three aspects of the roll-out (ailerons, rudder, and elevator).

____Climbs. To enter, use 3 steps:

____1. Ease the nose up to climb attitude.

____2. Increase power to climb value.

____3. Correct for "torque."

____Ensure that climb attitude is for V_y, the airspeed for maximum rate of climb. Check instruments; trim off control pressures. The rate of climb is the result of *excess* thrust horsepower existing at the pre-chosen airspeed and power setting. Thrust horsepower is that horsepower actually being used to fly the airplane.

____Straight and level (from a climb). Use 3 steps:

____1. Ease nose over to level flight, maintain climb power.

____2. Allow airspeed to build up, ease off right rudder.

____3. Adjust power to cruise setting as cruise airspeed is reached.

____Maintain the straight and level attitude; check altitude and heading and relieve pressure with trim.

____Glides (Descents).

      ____Carburetor heat ON or as a check for presence of carburetor ice, as recommended by the manufacturer.

      ____Close throttle smoothly.

      ____Hold airplane's nose in level flight attitude until airspeed reaches recommended glide value.

      ____Assume glide attitude; check airspeed indicator.

      ____Trim.

____Straight and level (from glide).

      ____Lead the selected altitude by about 50 feet for most trainers (it depends on the rate of descent).

      ____Add cruise power. Add enough power to stop descent, then adjust as necessary.

      ____Stop nose from rising with forward pressure of wheel or stick.

      ____Trim.

      ____Carburetor heat OFF (if applied).

      ____Maintain level flight attitude; check with instruments.

      ____Adjust power as needed.

____Climbing turns.

      ____Differences in left and right climbing turns ("torque" includes all left-turning tendencies of the airplane).

      ____Why the banks are kept shallow.

____Descending (gliding) turns.

      ____The airspeed is close to that of climb, but the rudder is comparatively weak because of a lack of slipstream (coordination problems).

      ____Banks should be kept shallow at this point of training.

■ **Flight Instruction: Dual 1.5 Hours**   (Including preflight check and cockpit checkout). The following should be discussed in the airplane on the ground or in flight, as applicable.

____Before takeoff.

      ____Preflight check. Use the manufacturer's checklist. Instructor should discuss the flight controls (*which* controls yaw, pitch, and roll) and *why* each item is checked on the airplane.

      ____Cockpit familiarization. Affirm that the student is comfortable and check seat adjustment to assure the ability to see out and move all controls. The sight-picture should be discussed.

      ____Instruments and controls including trim tabs. Briefly review the Four Fundamentals, with the student moving and watching the flight controls to simulate entry and recovery from each.

____Start engine (using the checklist) and explain the procedure. The instructor handles the tower communications.

____Taxi technique and brake system. Demonstrate and explain.

____Pretakeoff check, using checklist. Explain *why* for each item but don't go into detail.

____Takeoff. Make and explain the takeoff. The student's hand should be kept on the throttle during climb-out.

____Max rate-of-climb attitude. Demonstrate and explain.

____Level-off procedure. Demonstrate use of trim and leaning of mixture, noting that a full explanation will come later.

____Effects of controls.

____Student uses roll, pitch, and yaw controls (ailerons, elevators, and rudder), and also notes the effect of power changes.

____Turns. Demonstrate medium 30° bank and shallow turns with ailerons, rudder, and elevators. Student practices and checks for other traffic before turning. (Say, "Clear right, clear left"; use outside references.)

____Climb entry. Demonstrate and identify sight-picture attitude. Student practices. Note that right rudder is required to correct for "torque."

____Straight and level from a climb. Demonstrate and identify sight-picture attitude.

____Student climbs straight ahead and levels off.

____Student turns plane while climbing (with proper rudder), then levels off.

____Normal glide. Demonstrate and identify sight-picture attitude. Student practices setting up glide, then leveling off at a prechosen altitude.

____Return to airport. Instructor operates the radio and, with the student observing, goes through the checklist. Student flies and makes the turns in the pattern with instruction. Instructor handles power and flaps and takes over on final. Cover the use of the postflight checklist and explain postlanding, shutdown, and parking procedures.

____Postflight instruction.

____Evaluation.

____Critique.

____Review.

Assigned Reading.___*SPFM* Chapters 2-9 (Instructor: *FIM* Chapters 4 and 5)_____

Comments_____

_____

_____

Instructor_____

Date_____ Ground Instruction Time_____ Flight Time_____

Student Initials_____

UNIT 2 - Review and Repeat of Unit 1 (New material is added as necessary.)

■ **Ground Instruction. 1.0 Hour.** The following will be covered in the classroom or in the parked airplane as applicable:

____Review of Four Fundamentals.

____Airplane stability. The airplane wants to maintain equilibrium when trimmed. Briefly discuss non-technical longitudinal, lateral, and directional stability.

____Local traffic patterns and traffic patterns in general. Introduction of the wind indicators (wind sock, tetrahedron, wind tee). If the airport is controlled, the communications procedures and transponder should be briefly discussed. The instructor will handle the radio during this flight if at a controlled field, otherwise, the student will use Unicom or Multicom communications under supervision.

____Taxi (establish forward motion before attempting to turn). Check brakes as taxi begins. There is no emphasis on crosswind taxiing and control positions at this point.

    ____Brake system.

____Pretakeoff checklist.

    ____Magnetos check. Expected rpm drop; the ignition system.

    ____Carburetor heat check. Expected rpm drop; source of warm air.

    ____Ammeter and suction gauge. Cover their importance briefly.

    ____Idle rpm check.

____Use of trim.

____Wind drift correction theory. Introduction of the concept of crabbing and the rectangular course.

____Radio procedures to be used on this flight.

____Posttakeoff checklist.

____Climb straight ahead or as directed by ATC.

____At pattern altitude or as locally recommended, make a 45° left turn or climb straight out to exit airport traffic pattern (uncontrolled field).

____Electric fuel pump (if so equipped) is turned off at 1000 feet AGL or higher. Affirm by the fuel pressure gauge that the engine-driven pump is providing fuel pressure. Briefly explain use of electric fuel pump.

____Use shallow climbing turns to clear the area ahead.

____Review Four Fundamentals and effects of controls.

____Landing checklist.

____Descend to pattern altitude at least 2 miles out (non-tower airport). Do not descend into the pattern itself.

____Before entering the pattern:

    ____Check brake pedal(s) pressures.

____Fuel on fullest tank. Turn ON electric fuel pump, if so equipped. For Cessna C172 and C182, check fuel selector on BOTH; for C150/152 check that fuel selector is properly seated ON.

____Undercarriage coming down if not already welded in place.

____Mixture RICH. Short discussion of mixture control.

____Make traffic call.

____Enter downwind leg from 45° angle, if pattern calls for this.

____When established on downwind:

    ____Pull carb heat ON; leave ON for at least 10 seconds to clear out any ice that has formed. Leave either ON or OFF as required by manufacturer. Repeat why carburetor heat is used.

    ____Report position to tower or Unicom, as applicable.

    ____Perform G-U-M-P-P check. Reconfirm:

        ____Gas is on fullest tank.

        ____Undercarriage: check DOWN indications, if retractable.

        ____Mixture RICH.

        ____Pump: electric fuel pump ON, if so equipped, and POH indicates its use.

        ____Propeller: set pitch (if equipped).

____Abeam landing point, reduce power to approximately 1600 rpm (or to the particular training airplane requirements). Maintain altitude and let speed dissipate.

____When the airspeed is below V_{FE} (within white arc), lower flaps 2 notches (20° on C172/C152 and 25° on PA-28-161 Warrior). This procedure is to be modified to fit a particular airplane.

____Stabilize airspeed at recommended approach speed. Trim. Check prop again for go-around as recommended by the POH if the airplane has a constant-speed propeller.

____Student turns base (30° bank) and final (20° bank) as directed. Start the turn onto base when the touchdown area is 45° behind the wing. Set final landing flaps on base. This may be modified to setting landing flaps on final, depending on the airplane.

____Take over on final and land airplane, explaining procedure.

■ **Flight: Dual. 1.5 Hours**   (including preflight check).

____Review controls and instruments, with student seated in cockpit. Assure again that the student's seat is adjusted properly for height and fore and aft position.

____Preflight. Use checklist. Brief introduction to servicing airplane.

____Student uses prestart checklist, starts airplane, checks brakes as the airplane starts to move, and taxis to run-up position under instructor's supervision.

____Pretakeoff check. Use checklist.

    ____In addition to items on the checklist, explain that it is a good operating procedure to check idle RPM with carb heat ON or OFF (as the particular *POH* requires for approach and landing).

____Check traffic. Student uses radio under supervision.

____Student makes takeoff under supervision.

____Assume climb attitude; check airspeed indicator; trim.

____Explain scanning pattern when checking for other traffic. Use definite climbing turns to clear area.

____Climb to practice altitude.

____Level off, using the 3 steps.

____Turns, left and right (15-30° bank). Student practices.

____Glides. Student practices.

____Level off from the glide. Emphasize pitch-up effects of nose-up trim when the power is added during the recovery from the glide.

____Exercise. (Limits are ±100 feet altitude and ±10° heading in the straight and level portion.)

    ____Climb: 1000 feet.

    ____Straight and level: 2 minutes.

    ____Descend: 1000 feet.

    ____Straight and level: 2 minutes, then 180° turn.

    ____Repeat sequence.

____Demonstrate airplane longitudinal stability (phugoid oscillations–*SPFM*, Chapter 12; and *FIM*, Chapter 4).

____Demonstrate approach attitude with/without flaps, as applicable. (Refer to the POH for the airspeeds at various flap settings.)

____Student handles communications, under supervision, and enters traffic pattern, performs landing checklist, handles throttle and flaps, with help. Check brake pedal pressure. Instructor takes over on short final, and lands airplane while describing the process.

____Student taxis to ramp and shuts down, under supervision. (Instructor checks magneto ground wires for proper operation *at idle* by turning ignition OFF, then ON.) Student then runs up engine for a few seconds and shuts down the engine with the mixture, uses postlanding, shutdown, and postflight checklist, and secures the airplane. *Note: This is to be done for each flight without repeated instruction, but will be noted from time to time throughout this syllabus.*

____Postflight instruction.

    ____Evaluation.

    ____Critique.

    ____Review.

Assigned Reading.____*SPFM*, Chapters 10,12,13, and 21. *FIM*, Chapters 6 - 7._____

Comments_____

_____

Instructor_____

Date_____ Ground Instruction Time_____ Flight Time_____

Student Initials_____

UNIT 3 - Introduction of Steep Turns, Slow Flight, Rectangular Courses, and S-Turns Across a Road

■ Ground Instruction. 1.0 Hour

____Review Units 1 and 2 as necessary, with added emphasis on communications procedures at a controlled, or uncontrolled airport, as applicable. If controlled, briefly cover ATIS, clearance delivery, ground control, tower, departure and approach controls, and the frequencies required and *why* of each (*SPFM*, Chapter 21).

____Resolve questions relative to assigned reading material.

____Introduce 45°-banked power turns (360° and 720° turns) (*SPFM*, Chapter 10) and cover the following:

    ____Forces acting on airplane.

    ____Elevator, aileron, rudder, and power requirements.

    ____Pick a good reference on the horizon.

    ____Roll into 45°-banked turn (coordinated); then neutralize ailerons and rudder.

    ____Add power as bank increases.

    ____Add back pressure as bank increases.

    ____Continually check pitch and bank attitudes and check with altimeter.

    ____Roll out on original reference on horizon after one and two complete turns, as planned. As bank decreases, ease off back pressure and reduce power to cruise.

    ____Note correction for "torque" in left and right turns.

    ____The airplane may encounter its own wake turbulence.

____Introduce the Power required/Drag curve in simplified terms on chalkboard. (*SPFM*, Fig. 12-13).

    ____Induced drag (brief explanation).

    ____Parasite drag (brief explanation).

____Slow flight. Flight at minimum controllable airspeed (*SPFM*, Chapter 12; *FIM*, Chapter 6.) Slow flight is flight at such an airspeed that any further increase of angle of attack or decrease in airspeed would result in indications of an approaching stall.

    ____Decay of effectiveness of aileron, elevator, and rudder control at low airspeeds.

    ____The relationship of stall speed to angle of bank. With the airspeed close to stall at straight and level flight attitude, make all turns shallow because stall speed increases with bank, if back pressure is held.

    ____Control airspeed with pitch (elevators); altitude control with power (throttle).

    ____Note pitch changes necessary with addition or retraction of flaps during slow flight.

    ____Realistic distractions during slow flight. Scanning for other airplanes, etc.

    ____Straight and level altitude ±100 feet, heading ±10°.

____Review (if covered earlier) or introduce the concepts of wind correction.

____Procedures for flying over and parallel to a road.

____Rectangular course.

____S-turns across a road.

____Review radio and transponder procedures.

■ Flight: Dual. 1.0 Hour

____Student performs preflight, taxi, run-up, and takeoff, with supervision, and handles communications. Introduce flight control positions for various wind conditions.

____After takeoff, student assumes V_y attitude with attitude indicator covered; check the airspeed indicator.

____Climb to practice area and altitude.

____Demonstrate 45°-banked power turn for 720°.

____Student practices both right and left.

____Demonstrate slow flight, with shallow turns left and right.

____Student practices.

____Student establishes glide (carburetor heat as necessary) and levels off at approximately 600-1000 feet above the ground.

____Student flies over and also parallel to a road.

____Demonstrate rectangular course.

____Student practices.

____If time permits, demonstrate S-turns across a road.

____Student practices.

____Listen to ATC. Student calls approach control for return to airport (controlled field). Student calls Unicom or uses communications procedures as required (uncontrolled field).

____Student performs prelanding checklist, checks brake pedal pressure, enters normal traffic pattern, or as instructed by ATC, performs GUMPP check as a back-up; assumes approach attitude, and flies airplane to short final.

____Land the airplane, again describing the process.

____Student retracts flaps after clearing runway. Verify, as applicable, that the flap, not the gear handle, is being moved.

____Student taxis to ramp and performs shutdown and postflight checklist, under supervision.

____Postflight instruction.

    ____Evaluation.

    ____Critique.

    ____Review.

Assigned Reading. <u>Student: *SPFM*, Chapters 10-12, Review Chapters 4 -7.   Instructor: *FIM*, Chapters 9 and 11.</u>

Comments_____

_____

Instructor_____

Date _____ Ground Instruction Time_____ Flight Time_____

Student Initials_____

UNIT 4 - Introduction to Stalls

■ **Ground Instruction. 2.0 Hours.** May be done in two sessions.

____Review earlier flights as necessary.

____Resolve questions relating to assigned reading material, etc.

____Review use of mixture control.

____Review carburetion, carb ice, and carb heat.

____Discuss the following items relative to aerodynamics:

    ____Four Forces.

    ____Airfoils (wings and propellers).

    ____Chord.

    ____Angle of incidence.

    ____Relative wind.

    ____Angle of attack.

    ____Lift as it relates to angle of attack and airspeed: $L = C_L \, S \, ^{\rho/2} \, V^2$

    ____Critical angle of attack.

    ____More about flaps and why they are used.

    ____Stall as a function of angle of attack, not airspeed.

    ____The stall-warner leads the stall by a few knots (1-G stalls).

    ____Elevators as they relate to stall entry and recovery. To recover from a stall, point the airplane in the direction it's going.

    ____Power is used during a stall recovery *to minimize altitude loss.*

    ____To recover from the stall, simultaneously move the wheel or stick forward enough to break the stall (lower the angle of attack) and add full power to minimize altitude loss.

____Presolo stalls.

    ____Takeoff and departure stalls.

    ____Approach-to-landing stalls.

    ____Stalls as an aid to landings.

    ____Approach to, or partial (imminent), stalls.

    ____Recovery at the first indication of a buffet.

____Review concepts of wind correction.

____Review rectangular courses and S-turns as necessary, citing common errors.

____Turns about a point (optional discussion).

____Another look at the traffic pattern with added emphasis on pattern departures and entries.

____Normal approach and landing:

    ____Flap use.

    ____A stabilized approach is established by gradually reducing power, with elevator pressures trimmed off. Close throttle when runway is made.

    ____On final, at a height of approximately 20 feet (about hangar height), the transition is started from approach attitude to landing attitude. Gradually ease the wheel back. The transition must be gradual.

    ____About 19 of the 20 feet are used in the transition; the airplane is slowing up during this period.

    ____When the airplane is about 1 foot above the ground, it should be in the landing attitude–slightly nose high. The pilot will not be able to see directly over the nose in most airplanes. The pilot should look out along the left side of the nose, not over the nose, to line up with the runway. If the pilot can see over the nose, some airplanes won't have the proper landing attitude and will fly into the ground.

    ____Once in the landing attitude (about 1 foot above the ground):

        ____That attitude should be held. (The student should try to keep the airplane from land-ing.) As the airplane approaches the stall, back pressure on the wheel or stick is in-creased as necessary.

        ____Eyes should be focused about 10° left of the nose and far enough ahead so the ground is not blurred. Don't stare at one spot. Scan the surface.

        ____The airplane should be lined up with the runway and have wings level at touchdown (assuming no crosswind).

    ____During ground roll, back pressure should be held. The nosewheel should come down by itself.

    ____Brake as necessary. If the runway is long enough, the pilot may want to save brakes and turn off at a taxiway farther down the runway.

■ Flight: Dual. 1.0 Hour

____The student performs preflight, taxi, run-up, takeoff, and posttakeoff checklist. Student makes climbout at V_y, and handles communications including transponder, under supervision.

____Climb to practice area and altitude. Use clearing turns.

____Student practices 45°-banked power turns 360° left and right.

____Check 121.5 MHz for ELT transmissions every flight, then immediately return to primary fre-quency.

____Practice slow flight without flaps.

____While in slow flight, demonstrate simple stall and recovery without addition of power.

____Set up again for student to practice.

____Student establishes approach attitude with landing flaps. Demonstrate a full-stall landing at altitude.

____Student practices several, then cleans up airplane for climb.

____Student climbs at V_Y and levels off at a predetermined altitude.

____Demonstrate departure stall and recovery.

____Student practices departure stall.

____Demonstrate approach-to-landing stall (straight flight) and recovery.

____Student practices.

____Instructor demonstrates approaches to stalls.

____Student practices.

____Student executes descending turn(s) to an assigned heading and altitude for low work.

____Return to airport; listen to ATIS; check heading indicator with magnetic compass; call approach control (controlled airports). Call Unicom for airport advisory (uncontrolled airports).

____Check brake pedal pressure at this point in every flight.

____Student uses checklist; handles controls, using gradual reduction of power during the approach, and "lands" the airplane under supervision.

____Student retracts flaps after clearing runway (verifying that his hand is on the flap handle, not the gear handle).

____Student taxis to ramp and performs shutdown and postflight checklists.

____Postflight instruction.

    ____Evaluation.

    ____Critique.

    ____Review.

Assigned Reading.___Student: *SPFM*, Chapters 11, 12, 13 and 18. Instructor: *FIM*, Chapter 8; *POH*, Emergency Procedures.

Comments_____

_____

Date_____ Ground Instruction Time_____ Flight Time_____

Student Initials_____

UNIT 5 - Emergency Procedures

■ Ground Instruction. 1.0 Hour

____Review previous flights as necessary.

____Resolve questions relative to assigned reading material, etc.

____Wake turbulence and windshear avoidance procedures.

____Collision avoidance procedures.

____Briefly discuss maximum performance climbs, V_x and V_y.

____Review stall series.

____Introduction of emergency procedures. Use checklist.

    ____Simulated engine failure at cruise altitude. Close throttle, do not use mixture control.

        ____Carb heat ON as best glide is being set up.

        ____Establish best glide attitude and check with airspeed indicator. Trim to maintain.

        ____Pick a landing area and turn toward it.

        ____Mixture RICH.

        ____Electric fuel pump ON, and switch tanks, if so equipped.

        ____Confirm surface wind direction.

        ____After switching tanks, wait a few seconds to let the pump get fuel from new tank.

        ____The propeller will be windmilling; so if the fault (carb ice or a dry fuel tank) is remedied, the engine will start again. Do not try to use the starter.

____If engine starts, climb while circling the field until you are sure the problem is corrected before departing for further practice or return to the airport.

____If the engine will not restart, the following must be done before landing. Emphasize that these are all secondary to the primary need to keep the airplane under control at the best glide speed and make a safe landing.

    ____Declare an emergency on frequency 121.5, or if in contact with approach control or Unicom, stay on that frequency.

    ____Set transponder to 7700.

    ____Mixture to IDLE CUTOFF.

    ____Fuel selector OFF.

    ____Use flaps, as necessary, to land on chosen spot. It is best to use flaps in increments, but the first requirement is to make the field.

    ____After (electric) flaps have been extended and committed to landing, turn OFF master and ignition switches.

    ____Unlatch doors prior to touchdown.

    ____Land nose high, tail low; touch down as slowly as possible.

____THE FIRST RESPONSIBILITY IS TO CONTROL AND FLY THE AIRPLANE TO A SAFE LANDING. *Note that with low-altitude emergencies, line-of-sight factor (and distance), and time limits may preclude use of transponder or transmitter(s).*

____Engine running rough.

    ____Carb heat ON.

    ____Try individual mags.

    ____Mixture RICH.

    ____Fuel pump ON.

    ____Switch tanks.

    ____Primer locked?

    ____If problem is apparently solved, stay over "friendly" terrain and continue the lesson or return to the airport, depending on the problem.

    ____If immediate return to the airport is necessary, maintain altitude as long as safely practicable.

    ____Instructor introduces the concept of forced landing procedures while in descents and in the traffic pattern.

____Review the normal landing process for return to the home airport. Emphasize that the transition from approach attitude to landing attitude must be done carefully to avoid ballooning or flying into the ground.

■ Flight: Dual. 1.0 Hour

____Student performs all pretakeoff functions and executes the takeoff using appropriate checklists.

____Student climbs at V_y, establishes pitch attitude, then maintains that attitude with airspeed indicator (ASI) covered.

____Student levels off at practice altitude, then performs shallow (15°-bank) and medium (30°-bank) turns.

____Student does stall series (approach to and full stalls), in both straight and turning flight.

____Simulate power failure at altitude, by closing the throttle.

    ____Pull carb heat immediately to take advantage of engine's residual heat as the best glide airspeed is being set up.

    ____Electric fuel pump ON, switch tanks (if so equipped). Use the emergency checklist discussed during ground instruction.

    ____Pick a landing area. Check wind direction.

____Student glides toward landing spot. (Instructor takes over at safe altitude.)

____Set up rectangular course and/or S-turns across a road. Student practices.

____Return to the airport. Student handles the radio, flies the pattern, approach, and lands the airplane with supervision. Review the landing process during the return.

____The instructor may elect to have the student taxi back to shoot one or more patterns and full-stop landings before returning to the ramp.

____Student handles radio and flaps (flaps up after the airplane is off the runway) and taxis to the ramp. Student shuts down the airplane and uses postflight and shutdown checklist.

____Postflight instruction.

    ____Evaluation.

    ____Critique.

    ____Review.

Assigned Reading. _Student: SPFM,_ Chapters 12 and 13. Instructor: _FIM,_ Chapters 9 and 10 .____

Comments_____

_____

_____

Instructor_____

Date_____ Ground Instruction Time_____ Flight Time_____

Student Initials_____

UNIT 6 - Pattern Work - Normal Takeoffs and Landings and Introduction to Recoveries from Bad Situations

■ Ground Instruction. 1.0 Hour

____Review earlier Units as necessary.

____Resolve questions relative to assigned reading.

____Normal takeoff.

    ____Check for traffic. Do not blindly trust the tower at a controlled airport; emphasize that not all airplanes landing at an uncontrolled field have, or use, radios.

    ____Importance of lining up with center line of the runway.

    ____Smoothly apply full power, keep hand on throttle; emphasize that right rudder is needed to maintain runway alignment.

    ____Check engine instruments.

    ____Airspeed becomes "alive."

    ____As controls become firm, back pressure (tricycle gear) is used to obtain proper angle of attack on takeoff. Note that larger airplanes may use "numbers" for takeoff.

____Let the airplane fly itself off the runway; add right rudder as necessary as the nosewheel or tailwheel lifts off runway.

____Assume V_y attitude (unless V_x is needed); check with ASI.

____Climb out straight ahead; student's hand is on throttle; continue correcting for torque as necessary.

____Review traffic pattern (including departures and entries); prelanding checklist, both the initial and following power reduction; flap settings; establish approach attitude; and negotiate the approach down to the point of transition to a normal landing.

____Review landing process from breaking the glide to touchdown and roll out to taxi speed.

____The power-off approach.

____Introduction of the go-around or aborted landing procedure. Emphasize early decision and performance.

____Add full power to stop the descent. Forward pressure on wheel or stick as needed to keep nose from pitching up. Carburetor heat OFF if used.

____Trim.

____When there is a positive rate of climb, retract the flaps in 10° increments (or by notches, as applicable).

____Trim.

____Slight right turn then climb out parallel to runway (watch for traffic taking off).

____Trim.

____Landing gear retracted (if applicable).

____Discuss recoveries from bad situations during approach and the landing itself.

____Too high on final. Take it around.

____Too low on final. *Do not drag* the airplane up to the runway using minimum power (see *FIM*, Fig. 10-26).

____Flying into the ground (resulting in a bounce): Ease nose over. Add power *as the airplane descends* (if bounce is estimated 5 feet or more). If the bounce is higher, or if the runway is short, add full power and go around.

____Dropping it in. Add full power to lessen impact effects *as the airplane descends*.

■ Flight Instruction: Dual. 1.0 Hour

____Preflight.

____Taxi.

____Normal takeoff.

____Practice full-stop takeoff and landings.

____Demonstrate a power-off approach, explaining the difference in the downwind abeam position.

____Demonstrate an aborted approach/go-around.

    ____Student practices one or more as time allows.

____Postflight instruction.

    ____Evaluation.

    ____Critique.

    ____Review.

Assigned Reading.___Student: *SPFM,* Chapters 12, 13 and 21. Instructor: *FIM,* Chapter 10._____

Comments_____

_____

Instructor_____

Date_____Ground Instruction Time_____Flight Time_____

Student Initials_____

UNIT 7 - Takeoffs and Landings and a Review of Recoveries from Bad Situations

■ Ground Instruction. 1.0 Hour

____Review earlier flights as necessary.

____Resolve any questions relative to assigned reading.

____Traffic pattern, approach, and normal landing procedures.

____Introduce attitude flying in the pattern (loss of ASI).

____Communications requirements.

____Review recoveries from bad situations.

    ____Flying into the ground (bouncing).

    ____Dropping it in.

    ____Too low in the approach.

    ____Too high in the approach.

____Discuss emergencies on takeoff.

    ____Power failure while on runway.

    ____Power failure on climbout. Discuss the hazards of turning back.

    ____Partial power failure after lift-off.

____Door open during takeoff roll or after lift-off. Fly the airplane!

____Seat belt hanging outside. Describe sound. Fly the airplane!

____Emphasize that there are times when a takeoff must be aborted during the early part of the roll. Hold the wheel or stick full back while braking.

____Instructor briefly reviews forced landing procedures in cruise, descent and in the pattern.

____Introduce the crosswind takeoff.

____Taxiing in a strong crosswind. Flight control positions for various relative winds.

____Maintaining directional control in a strong crosswind (holding full ailerons into the wind and decreasing deflection as airspeed increases).

____Introduce the crosswind landing.

____Crabbing as a drift correction maneuver in the pattern.

____Sideslip as a drift correction maneuver on final and through touchdown.

____Crab approach and sideslip correction during the landing process.

____Introduce the effects of gusty air and wind gradients.

____Discuss loss of directional control; causal factors and recovery techniques.

____Review go-around procedure.

____From final approach.

____During landing flare.

____During turns to base or final.

____Clean up procedure.

____Cover wake turbulence avoidance procedures as necessary.

____Review scanning and collision avoidance procedures.

■ Flight: Dual. 1.0 Hour

____Student performs preflight check under supervision.

____Normal takeoffs and landings.

____Student climbs (at V_y) and sets up approach attitudes with ASI covered for at least two patterns.

____Intersperse full-stop and touch-and-go landings, as appropriate for the training situation.

____Simulate engine failure on climbout after takeoff and demonstrate procedure. (Get nose down immediately and make only shallow turns, depending on altitude.)

____Postflight instruction.

____Evaluation.

____Critique.

____Review.

Assigned Reading. Student: *SPFM*, Chapters 11, 12, 13 and 18. Instructor: *FIM*, Chapters 8 and 10.

Comments_____

_____

_____

Instructor_____

Date_____ Ground Instruction Time_____ Flight Time_____

Student Initials_____

UNIT 8 - Review of High-Altitude Emergencies; Introduction to Forward and Sideslips; Pattern Work at Another Airport

■ Ground Instruction. 1.0 Hour

____Review earlier flights as necessary.

____Resolve any questions relative to assigned reading material.

____Review simulated engine failure at altitude. This is a refresher only and will not be done in flight this Unit.

    ____Carburetor heat ON while establishing best glide speed.

    ____Picking a field.

    ____Restoring power if possible.

    ____Pattern procedures at field.

    ____Using voice or transponder procedures.

    ____Immediate prelanding steps.

    ____Postlanding procedures.

____Forward slip. As a means to lose altitude without picking up airspeed. Does the manufacturer allow slips with flaps? (Optional.) Note: Airplane and airport conditions may require that the student be proficient in forward slips before solo.

____Sideslip. As a means to correct for wind drift in a crosswind landing (if not covered in Unit 7).

____Crab approach and wing-down landing (if not covered in Unit 7).

____Landings with various flap settings (including no-flaps) in particular situations for later use.

____Review procedures for being too high or too low on final and for bounce or high level-off on landing.

____Steps in a go-around or aborted landing procedure.

____If time permits, draw the course to another airport on the sectional chart and briefly introduce the concept of pilotage.

____Note pattern altitudes, entry procedures, and radio communication requirements at the nearby practice airport.

____Traffic patterns, approaches, and landings applicable to varying environments of different airports.

■ Flight: Dual. 2.0 Hours

____Student executes preflight, run-up, and normal or crosswind takeoff. Crosswind work should be avoided if the student does not have a complete grasp of normal takeoffs and landings.

____Fly to another airport between 25 and 40 NM distance (if available) to practice takeoffs and landings.

____Review enroute to the practice airport:

    ____Climb attitudes and airspeeds at V_Y and V_X .

    ____Leveling-off procedures.

____If time permits, student practices 45°-banked power turns at altitude.

____Student descends to and enters the traffic pattern at the other airport for takeoffs and full-stop landings.

____Demonstrate procedures for too-high or too-low approaches, bounce recoveries, and throttle use if drop-in is imminent or occurring.

____Demonstrate go-around procedure.

____Set up a too-high and/or too-low approach and turn the airplane over to the student to check reaction and use of proper procedures.

____Student makes a go-around under direction.

____Crosswind takeoff and landing demonstration; practice if indicated, or not already practiced by airport and operating requirements.

____Return to home airport and shoot two or three landings (full-stop) as time and/or student fatigue permits.

____Postflight instruction.

    ____Evaluation.

    ____Critique.

    ____Review. Introduce briefly the required airplane papers and logbooks.

Assigned Reading. Student: *SPFM*, Chapters 12 and 13. Instructor: *FIM*, Chapter 10.

Comments_____

_____

Instructor_____

Date_____ Ground Instruction Time_____ Flight Time_____

Student Initials_____

UNIT 9 - Shooting Takeoffs and Landings

■ Ground Instruction. 1.0 Hour

____Review previous flights with emphasis on problem areas. Depending on the student's progress, the emergencies on takeoff and in the pattern should be reviewed. The emergencies should be reviewed a flight or two before solo but not emphasized in the briefing just before the solo flight, to avoid creating anxiety.

____Resolve questions concerning the traffic pattern: altitudes, departures, entries, etc.

____Review collision avoidance and wake turbulence.

____Briefly introduce cockpit resource management (CRM) and aeronautical decision making (ADM) concepts.

■ Flight: Dual. 1.0 Hour.
The last few flights should not require the instructor to say much, as noted in the text of *The Flight Instructor's Manual*, but the student's actions should be monitored. The instructor will be beginning get a sense of *when* the student is ready to solo. See the introductory paragraph at the beginning of this STAGE. It's best to shoot full-stop-and-taxi-back landings as the student approaches solo. This will make the transition to solo easier than if he has previously been shooting only touch-and-go's.

____Shoot full-stop landings.

____On at least one landing, set up a bounce condition from which the student must recover (if this hasn't already been practiced inadvertently during the previous takeoff and landing sessions). Also check the student's response to an imminent or actual drop-in.

____Student makes a go-around or aborted landing during the session.

____Depart and reenter the traffic pattern at least once during the period.

____Student makes *at least* one pattern with ASI *and* altimeter covered.

____Some instructors have the student depart the pattern and go to the practice area for stalls or slow flight to break up the continuous landing practice. (Students can lose progress by too much concentration on pattern work.)

____Postflight instruction.

____Evaluation.

____Critique.

____Review.

Assigned Reading. ___Student: *SPFM*, Chapter 13. Instructor: *FIM*, Chapter 10._____

Comments_____

_____

Instructor_____

Date_____ Ground Instruction Time_____ Flight Time_____

Student Initials_____

UNIT 10 - Presolo Knowledge Test—STAGE 1 Written Test. 1.0 Hour.

Before solo, the student must pass a knowledge test on FARs, operation of the airplane being used, and general safety practices. Make up the test based on the airplane, local conditions, and regulations and review it with the student after testing to assure a full understanding of missed questions. Some suggestions for the flight instructor:

I. THE AIRPLANE
A. *Fuel system.*
    1. Usable fuel (gallons).
    2. Weight of fuel per gallon.
    3. Expected average fuel consumption (gph) for cruise at the power setting used in the practice area.
    4. Minimum octane fuel to be used.
    5. Color of fuel used.
    6. Electrical auxiliary fuel pump operations, if applicable; when and why it is used.
    7. Preflight fuel check items and order of checking them.
B. *Oil system.*
    1. Brand and viscosity number of oil currently in use in the airplane.
    2. Total capacity and minimum operational volume (quarts).
C. *Electrical system.*
    1. Equipment and instruments (flaps, oil temperature, fuel gauges, etc.) that are electrically controlled.
    2. Effects of turning off the master (electrical) switch on the running of the engine in flight (none).
    3. Steps to be used in a suspected, or real, electrical fire.
D. *Instruments.*
    1. The instrument(s) that would be affected if the static port is totally closed.
    2. Location of the static port.
    3. Significance of each of the following airspeed markings:

  a. Green arc.

  b. White arc.

  c. Yellow arc.

  d. Red radial line.

  4. The max rate (V_y) and max angle (V_x) climb airspeeds. Note that these values change with density-altitude but have the student give the numbers for sea level.

  5. The number of seconds for the oil pressure to come up to normal in moderate ambient temperatures.

  6. The normal fuel pressure (if applicable), in psi (green arc)

  7. Steps to follow if the low-voltage light comes on in flight. Discuss several options with the student. If in the traffic pattern, continue the approach and land, terminating the flight. If some distance from the airport, recycle the master (electrical) switch, etc.

E. *Airplane papers.* (The flight instructor will have introduced the airplane papers during one or more of the presolo sessions.)

  1. Airworthiness certificate. Must be displayed. For how long is it valid?

  2. Registration certificate. When must it be replaced?

  3. Weight and balance information (*SPFM*, Chapter 23).

  4. Logbooks for airframe, engine, and propeller. Annual and 100-hour inspections.

  5. Operating limitations. Performance charts and how to use them for finding required runway lengths.

F. *Emergency procedures and equipment malfunctions.*

  1. Approaches to a landing area with an engine malfunction.

  2. Loss of airspeed indicator.

  3. Flaps inoperative.

  4. Rough-running engine—symptoms.

    a. Magneto problems.

    b. Carburetor ice.

    c. Fuel starvation.

II. TRAFFIC PATTERN AND PRACTICE AREA OPERATIONS

A. *Pattern.*

  1. The student should sketch the traffic pattern for the home airport for all runways. (If some patterns are right hand, this should be shown.) The altitudes (MSL) for the first turn after takeoff and the downwind leg(s) for normal approach should be shown.

  2. The student should sketch the approximate practice area showing towns and other landmarks that would help pin down the boundaries.

  3. The student should list airports in the area that can be used as alternates should weather or other factors preclude the return to the home base. (The student may have already shot landings at one or more of these airports, but should review them here.)

B. *Discuss the following situation with the student:* You are solo and doing a takeoff and departure stall in a right-climbing turn at the break and the airplane rolls abruptly to the left. The nose is down and the airplane is rotating. The wheel is full back but is not bringing the nose up. You should immediately do which of the following?

  1. Hold the control wheel full back and make sure the engine is at full power to try to stop the altitude loss.

  2. Close the throttle, apply full rudder opposite to the rotation, and immediately move the control wheel forward. (This is the correct answer for a spin recovery, but the *POH* procedure should be the determining factor.) Briefly discuss spins.

  3. Apply rudder into (with) the rotation, maintain full power and stop the nose from getting farther down with full-back wheel.

III. AIRSPACE AND REGULATIONS

A. *Airspace.* The instructor should be sure that the student is aware of airspace that would be a factor in solo operations in the pattern and practice area. The instructor should cover local airways and Class B, C, D, E, and G airspace; prohibited and restricted areas; plus Military Operations Areas (MOAs) and Military Training Routes, *but only* as they would affect the student at this stage of his local training. To ask about operations in Class B airspace 200 miles away does nothing but clutter up the student's mind with trivia.

B. *Regulations.*

1. The instructor's questions on Regulations Parts 61 and 91 should also be applicable to the current situation. For instance, the requirement of 30 minutes minimum fuel reserve for VFR flight (FAR 91.151) has probably been superseded by your, or the school's, hour or more fuel reserve requirement.

2. The instructor should apply questions to what is required at this point in his training, both general and local. For instance, there may be local restrictions such as, "Don't fly below 3000 feet AGL over the Veterans' Hospital" and other rules that would only apply to your operations.

3. While it's not required, the instructor might later want to make up a written test for the student to take before going on that first solo cross-country. Ask general questions on FARs and airspace; the trip itself, with emphasis on the destination airport facilities, traffic patterns, etc.; plus alternate plans for weather and other factors. The instructor might not want to bother with tests for each of the following solo cross-countries if he or she is satisfied with the student's judgment on the first one.

4. The instructor should keep the presolo written test and any others on file for future reference for at least 3 years (FAR 61.189).

5. To repeat: The point of the instructor making up test(s) rather than getting a "canned" quiz from commercial publishers is twofold: First, he or she knows more about local operations and requirements than somebody sitting at a desk a thousand miles away. Second, making up such a test will add knowledge of the requirements for a safe postsolo period for the student and will also help gain experience in this area for the instructor's groundschool instructing.

UNIT 11 - First Solo

■ **Flight: Dual/Solo. 1.0 Hour.**   This Unit, while numbered 11, will occur for most students between the 12th and 20th Hours, depending on the student and frequency of flights. Be sure that the presolo knowledge test has been completed, graded, and reviewed and is on file; the following should be checked:

____All areas of the presolo syllabus have been covered.

____Any special, local course rules and procedures are clear to the student.

____Emergencies and go-around procedures are clear to the student.

____The wind and weather conditions are good for a pattern solo for this particular student.

____The student will shoot three full-stop landings unless special circumstances assure that touch-and-go's would be better for this solo. Full-stop landings are best for the first solo (*FIM*, Chapter 10).

____Air traffic is such that no conflict should occur while the student is in the pattern.

____The student's pilot/medical certificate is in order and is signed before the solo takes place.

____Logbook endorsement.

■ **Endorsement for Presolo Flight Training: FAR 61.87**

I have given Mr/Ms _____ the flight instruction required by FAR 61.87 in a (*make and model aircraft*). He/She has demonstrated proficiency in the applicable maneuvers and procedures listed in FAR 61.87[d] (1 through 15) and is competent to make solo flights in a (*make and model aircraft*).

Signed_____ Date_____ Cert. No._____ Exp._____

Debriefing: (Fill out logbook.)

Number of takeoffs and landings_____ Dual_____ Solo_____

Comments _____

_____

Date_____ Instructor_____ Cert. No._____ Exp. Date_____

Student Initials_____

UNIT 12 - Summary

As noted earlier, the 10th and following Units until solo are essentially a review of the previous instruction, with the instructor doing less flying and less talking as the student becomes more proficient. Again, most students solo between the 12th and 20th Hours, depending on consistency of schedule, airplane type, and airport environment (traffic density, tower control, crosswinds, length and width of runways, and other factors).

To repeat, the term "Unit" rather than "Flight" has been used in the syllabus because the items in a Unit may need more than one flight to complete. For instance, on the third flight the student may still need work on the Four Fundamentals in the first and second units. A timid student shouldn't be introduced to emergency procedures (in Unit 5) as early as the fifth flight; he's still worried about banking over 10° at that point. Certainly an instructor would not want to get into crosswind takeoffs and landings if the student doesn't understand the no-wind or wind-down-the-runway techniques. Under certain conditions, a delay until after solo for the student to begin crosswind procedures, or slips, may be wise, but the instructor should discuss and demonstrate these maneuvers before solo (see Units 7 and 8).

While the suggested order of introduction is listed here, it may need to be modified for a particular student and/or local requirements.

STAGE 2

IMMEDIATE POSTSOLO

For many student pilots, the first solo is the high point, something that has been looked forward to for the past several hours of flying, but a letdown is often experienced in the following hours. Most drop-outs occur in STAGE 2.

The immediate postsolo syllabus material is intended to keep up the student's interest and to continue progress toward achieving the experience and skills required to safely exercise the private certificate. Early in STAGE 2 and throughout, work in dual flights to cover crosswind takeoffs and landings for the student when local conditions allow.

UNIT 1 - Postsolo Flight Check and Supervised Solo

■ **Ground Instruction. 0.5 Hour**

____Review first solo flight.

____Review traffic pattern and collision avoidance.

____Review traffic communications.

____Review go-arounds, bounce recoveries, being too high or too low on final, and crosswind techniques.

____Plan at least three dual takeoffs and landings followed by solo patterns as time and weather permit.

■ **Flight: Dual. 1.0 Hour.**   Pick the time, weather and wind conditions. The conditions for any crosswind operation during the solo portion of this Unit should be carefully considered.

____Preflight inspection.

____Start and taxi.

____Run-up.

____Takeoff, and depart the pattern for the practice area.

____Check the student on the following maneuvers by the chief, or another, instructor:

    ____720° power turns.

    ____Slow flight.

    ____Stalls in all configurations.

    ____Simulated high-altitude emergency.

___Rectangular course.

___S-turns across the road.

___Introduce turns around a point (demonstration).

___Return and traffic pattern entry.

___Crosswind takeoff and landing instruction as conditions allow.

___Three takeoffs, traffic pattern and landings, with review and instruction as necessary for safe solo landings.

■ **Flight: Supervised Solo. 1.0 Hour.**   Following the postsolo flight check, the student should shoot at least six solo takeoffs and landings to a full stop at the home airport. Touch-and-go's are not recommended for this flight. The flight instructor should be present at the airport during this period, preferably at a position where critiques of each pattern may be made. A table, such as given in Figure 10-39, *The Flight Instructor's Manual*, should be used for critique. The following items should be noted for the critique:

___Takeoff.

___Climb.

___Crosswind leg.

___Downwind leg.

___Base leg.

___Final.

___Forward slips to landings if the student has had dual in the procedure during the presolo stage.

___Go-arounds (as applicable).

___Roundout or flare.

___Touchdown.

___Rollout.

___Crosswind corrections (as applicable).

___Taxi.

___Parking.

___Shutdown.

___Postflight check.

___Postflight instruction

    ___Evaluation.

    ___Critique.

    ___Review.

Assigned Reading.___Student: *SPFM*, Chapters 13, 18 and 21. Instructor: *FIM*, Chapter 10.___

Comments_____

_____

Instructor_____

Date_____ Ground Instruction Time_____ Flight Time_____

Student Initials_____

UNIT 2 - Solo in the Pattern

■ **Flight: Solo. 1.0 Hour.**   No ground instruction for this Unit, unless the student has questions. If the first solo takeoffs, patterns, and landings are satisfactory, the student should be scheduled for an hour in the pattern under selected wind and weather conditions. The instructor is aware of the time, place, and weather conditions of this flight and has briefed the student on the time in the pattern. Depending on the student, full-stop landings may be interspersed with touch-and-go's as traffic and communications allow.

____Postflight instruction.

____Evaluation.

____Critique.

____Review.

Assigned Reading. Student: *SPFM*, Chapters 10,12 and 13. Instructor: *FIM*, Chapters 6, 7 and 10

Comments_____

_____

Instructor_____

Date_____ Ground Instruction Time_____ Flight Time_____

Student Initials_____

UNIT 3 - Review of Stalls, Introduction to Incipient Spins and Recoveries

Before the student is sent out to practice solo slow flight and stalls he or she should have demonstrated an understanding of stalls and incipient spin recognition and recovery.

■ **Ground Instruction. 1.0 Hour.** Discuss the following:

____Stalls.

    ____Slow flight.

    ____Theory of stalls. *Why* the stall occurs.

    ____Stalls as a function of angle of attack, not airspeed.

    ____Accelerated stalls.

____Review of straight-ahead and turning stalls introduced before solo. (Takeoff and departure, approach-to-landing stalls clean and dirty).

____Stall recoveries: at first indication and when stall has broken.

____Approach to stalls: recognition and recovery.

____Spins.

    ____Spin theory. *Why* an airplane spins.

    ____Autorotation and the incipient spin.

    ____Developed spin.

____Spin recovery procedure (refer to *POH*) but cover generally:

    ____Close the throttle.

    ____Neutralize the ailerons.

    ____Ensure that the flaps are up.

    ____Opposite rudder. Discuss different methods for detecting rotation direction.

    ____Brisk forward wheel or stick applied after rudder has reached stop.

    ____When rotation stops, neutralize the rudder to avoid a progressive (opposite) spin.

    ____Pull out of dive.

    ____Immediate response necessary when high angle of attack and yaw are occurring.

    ____Neutralize rudder and ailerons.

    ____Wheel forward if spin is impending and confusion exists as to direction (to avoid using the wrong rudder).

    ____No solo spin practice!

    ____Cross-control (skidding) stall. Where and when it is most likely to happen. If the situation is pending on the turn to final, get off the rudder, relax the excessive back pressure, the wings should be rolled level and a safe go-around executed.

____Review of high-altitude emergencies.

____Return to airport. Student handles all communications.

■ Flight Instruction: Dual. 1.0 Hour

____Preflight, start, and taxi.

____Use of checklist.

____Taxi, takeoff, climb, and flight to practice area.

____Practice takeoff and departure and approach-to-landing stalls. Recover after the stall has broken.

____Approaches to stalls and recoveries at the first indication of impending stall.

____Incipient spin entries and recoveries, left and right. Assure that FARs 91.303 and 91.307 are respected, and the airplane is certified for spins.

____Demonstrate cross-control (skidded) stalls.

____Demonstrate accelerated stalls (depending on airplane and student).

____High-altitude emergency. Instructor points out suitable areas for landing and takes over at safe altitude.

____Return to airport. Student handles all communications.

This Unit is an "active" flight, so that frequent breaks should be made for the student's comfort. There should be an open "comfort bag" under the instructor's seat – just in case.

____Postflight instruction.

____Evaluation.

____Critique.

____Review.

Assigned Reading. Student: *SPFM*, Chapters 10, 12, and 13. Instructor: *FIM*, Chapters 6, 7, and 9.

Comments_____

_____

Instructor_____

Date_____ Ground Instruction Time_____ Flight Time_____

Student Initials_____

UNIT 4 - Solo in the Practice Area and Pattern I

■ **Ground Instruction. 1.0 Hour.** The following should be reviewed:

____Pattern departure and entry.

____Practice area.

____Collision avoidance procedures.

____Steep turns.

____Descents. Use of carburetor heat and suggested airspeeds.

____S-turns.

____Turns around a point.

____Return to traffic pattern.

____Communications procedures.

■ **Flight: Solo. 1.0 Hour**

____Climb to altitude and complete a 720° power turn (45° bank) in each direction.

____Slow flight.

____Descend to the altitude for rectangular course in practice area (use carb heat). NO SIMU-
LATED EMERGENCIES SOLO!

____S-turns.

____Turns around a point.

____Return to traffic pattern and shoot at least three takeoffs and landings (full stop preferred). *No
stall practice on this flight.*

____Postflight instruction.

    ____Evaluation.

    ____Critique.

    ____Review.

Assigned Reading.___Student: *SPFM*, Chapters 12 and 14. Instructor: *FIM*, Chapters 6, 11, and 21.___

Comments_____

_____

Instructor_____

Date_____ Ground Instruction Time_____ Flight Time_____

Student Initials_____

UNIT 5 - Solo in the Practice Area and Pattern II

■ **Ground Instruction. 0.5 Hour.**   A review and briefing for solo practice of steep turns, stalls, slow flight, wind drift correction maneuvers, and takeoffs and landings.

■ **Flight: Solo. 1.5 Hours**

As many of the following maneuvers should be done as time allows.

____Steep turns (45°-banked 720° power turns). PTS (Practical Test Standards) limits are 45°-banked ±5°, airspeed ±10 knots, altitude ±100 feet, and rollout on the entry heading ±10°.

____Slow flight. Maintain the specified altitude ±100 feet, specified heading ±10° and specified airspeed +10/-5 knots.

____Stalls; straight ahead and turning stalls, power-on and power-off (clean).

____Rectangular course. Maintain altitude ±100 feet, airspeed ±10 knots. Look around!

____S-turns. Altitude ±100 feet, airspeed ±10 knots.

____Turns around a point. Fly turns in both directions; exit at the same altitude and heading as the entry.

____Normal or crosswind takeoffs and landings.

____Review flight with instructor.

____Postflight instruction.

      ____Evaluation.

      ____Critique.

      ____Review.

Assigned Reading.   Student: *SPFM*, Chapters 3 and 15. Instructor: *FIM*, Chapters 12 and 24.____

Comments_____

_____

Instructor_____

Date_____ Ground Instruction Time_____ Flight Time_____

Student Initials_____

UNIT 6 - Introduction to Emergency Instrument Flying

■ Ground Instruction. 1.0 Hour

____Answer any questions that may have arisen since the solo flights in the practice area and pattern.

____Introduction to flying by instruments.

    ____Aviate, navigate, communicate.

    ____The pitot-static instruments (briefly).

    ____The gyro instruments (briefly).

    ____Instrument scan.

    ____Bank instruments.

    ____Pitch instruments.

    ____Control and performance instruments.

    ____Attitude plus power equals performance.

    ____Primary and supporting instruments.

____Four Fundamentals and the instrument indications.

    ____Straight and level.

    ____Climbs.

    ____Descents.

    ____Turns.

■ Flight: Dual. 1.0 Hour.

The student will do all flying except when instructor demonstration may be required. This Unit may use a flight simulator or flight training device (and authorized instructor) for FARs 61.4 and 61.109.

____Preflight with added checks on instrument factors (antennas, pitot heat, and other IFR equipment). No detailed discussion at this point.

____Start.

____Taxi.

____Pretakeoff check.

____Takeoff.

____Climb, referring to flight instruments, not hooded.

____Level off, power and trim.

____Straight and level using all flight instruments but not hooded.

____Straight and level, hooded.

____Climbs, using all instruments, hooded.

____Turns, using all instruments, not hooded.

____Turns to headings, hooded.

____Straight descents, all instruments, not hooded.

____Straight descents to a prechosen altitude, hooded.

____Alternating straight climbs, straight and level, and descents, hooded.

____Constant altitude turns to prechosen headings, hooded.

____Break from hood work for 5 minutes or more.

____Descend to traffic pattern altitude, hooded, if the student is not fatigued.

____Visual pattern entry and landing: emphasize use of prelanding checklist.

____Shutdown procedures.

____Postflight instruction.

    ____Evaluation

    ____Critique

    ____Review

Assigned Reading.   Student: *SPFM*, Chapters 10, 15, 16 and 18. Instructor: *FIM*, Chapters 6-12.

Comments_____

_____

Instructor_____

Date_____ Ground Instruction Time_____ Flight Time_____

Student Initials_____

UNIT 7 - DUAL REVIEW OF PREVIOUS INSTRUCTION

■ Ground Instruction. 1.0 Hour

____Review stalls, slow flight, and spin recoveries.

____Review Four Fundamentals and instrument flying.

____Review high-altitude emergencies.

____Review rectangular course and S-turns.

____Review turns around a point.

____Review crosswind takeoff and landing techniques.

■ Flight: Dual. 1.0 Hour

____Preflight, start, and taxi.

____Use of check list.

____Taxi, takeoff, climb, and flight to practice area.

____Four Fundamentals, using all instruments, hooded.

____High-altitude emergency, not hooded, naturally.

____S-Turns and turns around a point.

____Return to airport. Student handles communications and enters traffic pattern by approach control/tower clearance, or Unicom, as applicable.

____Student shoots crosswind takeoff(s) and landing(s) if conditions allow.

____Demonstrate no-flaps landing.

____Student shoots no-flaps landing.

____Postflight shutdown and check.

____Postflight instruction.

    ____Evaluation.

    ____Critique.

    ____Review.

Assigned Reading.   Student: *SPFM*, Chapters 10,12, and 13. Instructor: *FIM* ,Chapters 6 and 7.

Comments_____

_____

Instructor_____

Date_____ Ground Instruction Time_____ Flight Time_____

Student Initials_____

UNIT 8 - Solo in the Practice Area and Pattern III

■ Ground Instruction. 0.5 Hour

____Review of preceding dual flight.

____Brief on slow fight and stalls.

____Stalls, clean and dirty. (Only takeoff and departure and approach to landing stalls to be practiced on this flight.)

____Steep turns (720° power turns at 45° banks).

____Emphasis on clearing the area and collision avoidance procedures.

____Review of wake turbulence avoidance.

____Rectangular course and turns around a point

■ Flight: Solo. 1.0 Hour

____Slow flight and stalls.

____720° power turns.

____Descend to practice rectangular course, S-turns, and turns around a point.

____Return to airport. Shoot at least three takeoffs and landings (full stop).

____Postflight instruction.

    ____Evaluation.

    ____Critique.

    ____Review.

Assigned Reading. Student: *SPFM*, Study Chapter 17; review Chapters 13, 15, 16, and 18. Instructor: *FIM*, Chapter 11.

Comments_____

_____

Instructor_____

Date_____ Ground Instruction Time_____ Flight Time_____

Student Initials_____

UNIT 9 - Dual: Short- and Soft-Field Takeoffs and Landings

■ Ground Instruction. 1.0 Hour

____Dragging the area for an off-airport landing (emergency only).

____Short-field takeoffs.

____Max angle and obstacle-clearance climbs.

____Soft-field takeoff techniques.

____Soft-field postliftoff procedures.

____Comparison of short- and soft-field takeoff configurations and procedures.

____Short-field approach and landings; airspeed control, use of power, 1.3 V_{so} approach speed.

____Braking techniques.

____Soft-field approach and landing techniques.

____Comparison of short- and soft-field landing configurations and procedures.

____Crosswind takeoff and landing review.

____High density-altitude takeoffs (restricted RPM to simulate conditions).

■ Flight: Dual. 1.5 Hours

____Student preflights, starts, performs pretakeoff check and takeoff.

____Climb to practice area and altitude; emphasize clearing turns in climb.

____Steep turns.

____Slow flight.

____High-altitude emergency.

____Instructor demonstrates dragging the area at a field in the practice area.

____Return to airport; pattern entry. Emphasize airspeed and altitude control on approach.

____Instructor demonstrates at least one short-field takeoff and climb.

____Instructor demonstrates at least one short-field approach and landing.

____Student does at least one short-field takeoff and landing.

____Instructor demonstrates at least one soft-field approach and landing.

____Student performs at least one soft-field approach and landing.

____Instructor has student perform one or more high density-altitude takeoffs (restricted RPM).

____Postflight instruction.

    ____Evaluation.

    ____Critique.

    ____Review.

The instructor may elect to follow this flight with another dual on the special takeoff and landing techniques before having the student practice them solo.

Assigned Reading. Student: *SPFM*, Chapters 13 and 17. Instructor: *FIM*, Chapters 6, 7, 9, 10, and 11.

Comments_____

_____

Instructor_____

Date_____ Ground Instruction Time_____ Flight Time_____

Student Initials_____

UNIT 10 - Solo in the Practice Area and Pattern IV

This Unit will combine Parts I, II, and III of "Solo in Practice Area and in the Pattern", plus short- and soft-field and crosswind takeoffs and landings (dual in Unit 9).

■ **Ground Instruction. 0.5 Hour.** The instructor will review special takeoff and landing techniques with the student and may elect to give another Unit of dual in these procedures before solo work.

■ **Flight: Solo. 1.5 Hours**

____Short-field takeoff and obstacle climb. Use proper flap configuration and obstacle-clearance airspeed, or V_x, as required.

____Steep turns 45° bank ±5°, airspeed ±10 knots, altitude ±100 feet; rollout on the entry heading ±10°.

____Slow flight. Altitude ±100 feet, heading ±10°, and airspeed +10/-5 knots.

____Stalls. Takeoff and departure and approach-to-landing stalls.

____Rectangular course, ±100 feet, ±10 knots.

____S-turns, ±100 feet, ±10 knots.

____Turns around a point; turns in both directions, exit at the same altitude and heading as the entry, ±100 feet, ±10 knots.

____Return to airport using proper communications, pattern entry, and pattern.

____Short- and soft-field takeoffs and landings to a full stop. No touch-and-go's.

____Instructor should fill out critique sheet (*FIM*, Fig. 10-39) if takeoffs and landings were observed.

____Review flight with instructor.

Assigned reading. <u>Student: *SPFM*, Chapters 12 and 15. Instructor: *FIM*, Chapter 12.</u>

Comments_____

_____

Instructor_____

Date_____Ground Instruction Time_____Flight Time_____

Student Initials_____

UNIT 11 - Emergency Instrument Flying

■ **Ground Instruction. 1.0 Hour**

____Review Four Fundamentals.

    ____Straight and Level.

    ____Climbs.

    ____Glides (descents).

    ____Turns.

____Six Basic Maneuvers.

    ____Straight and level ± 10° of heading.

    ____Left and right 180° constant-altitude turns ±10° on rollout.

    ____Shallow, climbing, constant-airspeed turns to a predetermined altitude and heading.

    ____Shallow, descending turns to a predetermined altitude and heading.

    ____Recovery from the start of a power-on spiral.

    ____Recovery from the approach to a climbing stall.

■ **Flight: Dual. 1.5 Hours.** This Unit may use a flight simulator or flight training device (and authorized instructor) for FARs 61.4 and 61.109.

____Takeoff and climb to the practice area and safe altitude.

____Four Fundamentals, hooded.

____Six Basic Maneuvers, hooded.

    ____Straight and level ± 10° of heading.

    ____Left and right 180° constant-altitude turns ± 10° on rollout.

    ____Shallow, climbing, constant-airspeed turns to a predetermined altitude and heading.

_____Shallow, descending turns to a predetermined altitude and heading.

_____Recovery from the start of a power-on spiral.

_____Recovery from the approach to a climbing stall.

_____Introduction to hooded tracking with VOR or homing to an NDB (if equipment is available).

_____Return to vicinity of airport. Student flying, hooded, with instructor acting as radar.

_____Simulated ASR or PAR if traffic permits.

_____Short- or soft-field approach and landing, visual!

_____Postflight instruction.

_____Evaluation.

_____Critique.

_____Review.

Assigned Reading. Student: _SPFM_, Chapters 13 and 17. Instructor: _FIM_, Chapter 10.

Comments_____

_____

Instructor_____

Date_____ Ground Instruction Time_____ Flight Time_____

Student Initials_____

UNIT 12 - Solo Takeoffs and Landings in the Local Pattern

■ **Ground Instruction. 1.0 Hour.**   The instructor should review normal, crosswind, soft- and short-field takeoffs and landings and assure that the student has a full understanding of each of the requirements. A check and confirmation of the strength of a crosswind component should be made before sending the student solo; this is particularly important for short- and soft-field takeoffs and landings. The student may leave and reenter the pattern for practice and a "break," traffic permitting.

■ **Flight: Solo. 1.0 Hour**

____Preflight check.

____Special takeoff and landing procedures.

    ____Short-field takeoffs and landings.

    ____Soft-field takeoffs and landings.

    ____Crosswind takeoffs and landings.

    ____Slips to landing, if approved.

____Debriefing by flight instructor (review of flight with student).

    ____Evaluation.

    ____Critique.

    ____Review.

Assigned Reading. Student: *SPFM*, Chapters 19 - 25; in preparation for STAGE 3 Cross-Country Flying.

Comments_____

_____

Instructor_____

Date_____ Ground Instruction Time_____ Flight Time_____

Student Initials_____

UNIT 13 - STAGE 2 Written Test

The student will be given a multiple-choice test made up by the flight school or instructor to cover the ground and flight instruction given in STAGE 2. The questions should cover at least the following areas:

____Preflight inspection, possible problem areas.

____Pretakeoff checks, possible problem areas.

____Traffic pattern and communications.

____Go-around procedures.

____Bounce recoveries.

____Crosswind takeoff and landing techniques.

____Turns around a point.

____Flight by reference to instruments.

    ____The Six Basic Maneuvers.

____Takeoff and departure stalls.

____Approach-to-landing stalls.

____Slow flight.

____High-altitude emergencies.

____Airplane "numbers." Max rate and max angle climb airspeeds, max distance glide, etc.

____Wake turbulence avoidance.

____Collision avoidance.

____Soft- and short-field takeoffs and landings.

____Postlanding procedures.

Grade_____

Instructor_____

Date_____Ground Instruction Time_____

Student Initials_____

STAGE 3

CROSS-COUNTRY

This STAGE introduces and completes the dual and solo cross-country requirements for the private certificate for FAR Parts 61 or 141. If possible, a dual flight to a public-use airport with a grass runway should be accomplished sometime during the course so that the student can have experience taking off and landing on a non-paved surface.

UNIT 1 - Solo Flight to a Nearby Airport

■ **Ground Instruction. 2.0 Hours.** Note that this is an introduction to "real" cross-country. The following items should be covered:

____Brief for a solo flight to another (non-tower) airport within 25 NM, or more than 25 NM but not more than 50NM, if the flight instructor and student have flown to that airport and the student is competent and proficient to enter and depart the traffic pattern and to make those landings or takeoffs. The student pilot's logbook must be endorsed with an authorization for one or more flights (FAR 61.93[b]). See STAGE 1, Unit 8.

____Ground servicing, including fuel and oil.

____Use of external power (as available) for starting.

____Sectional chart, use of plotter (courses and distance) to and from the airport previously flown to dual at a distance of 40 NM (as an example).

____Introduction to weather and other services available for cross-country flights.

    ____Hourly weather reports (METARs).

    ____Terminal forecasts (TAFs).

    ____Area forecasts.

    ____Winds aloft forecasts.

    ____Pilot reports (PIREPs).

    ____NOTAMs.

____Check weather. Suggested ceiling/visibility minimums of 6000 feet/8 statute miles for solo flight with a forecast of this minimum or better for the next 4 to 6 hours. Student calls FSS or AFSS with flight instructor on the phone extension (or speaker) so that a student-instructor conference may be made during and/or after the briefing. Check surface winds at destination airport.

____Alternate plans available in case of weather change on outbound and inbound legs.

____Review student's paperwork and sign the logbook.

■ **Flight: Solo. 2.0 Hours**

____Student does normal full-stop takeoffs and landings at the other airport.

____Postflight debriefing.

    ____Review solo flight to the other airport; ask and answer questions and discuss any problems the student may have encountered enroute.

    ____Evaluation

    ____Critique

    ____Review

Assigned Reading.   Student: *SPFM*, Chapters 19-21. Instructor: *FIM*, Chapters 13, 14 and 25.

Comments_____

_____

Instructor_____

Date_____ Ground Instruction Time_____ Flight Time_____

Student Initials_____

UNIT 2 - Use of Radio Aids to Navigation and Communications Procedures

■ **Ground Instruction. 3.0 Hours** (may be done in two or three sessions).   It is suggested that this Unit be taught at a formal groundschool or one-on-one, with no plans to fly immediately following. Certain parts will be repeated as part of the preflight briefings in this STAGE, but the student should have a thorough introduction to communications and navigation aids available early on. Some of this Unit will have been covered, earlier as needed, but should be "pulled together" here.

____VOR VHF, Omni-Directional Range.

    ____The major components (omnibearing selector [OBS], course deviation indicator [CDI], and TO-FROM indicator).

    ____Theory of how the VOR works (*SPFM*, Fig. 21-4).

    ____Frequencies in use. Compare with communications frequencies already used.

    ____Advantages (VHF, OMNI - selection of bearings).

    ____Disadvantage, line of sight - (*SPFM*, Fig. 21-8).

    ____Selecting the proper frequency (identify!) and setting up the VOR receiver to track TO or FROM a station.

    ____Discussion of tracking directly TO or FROM the station.

    ____180° ambiguity. The CDI needle reads in reverse if the OBS is set opposite to the desired course.

____Brief mention of tracking to the station on a desired bearing but emphasize, at this point, that tracking directly to a VOR is the recommended procedure if lost or "uncertain."

____Principles of locating by cross-bearing (*SPFM*, Fig. 25-4).

____Station crossing. Indications.

____Airplane may not always cross directly over the station; the needle may be slightly to one side or the other for an acceptable station passage.

____NDB, non-directional beacon.

____Components of the automatic direction finder equipment in the airplane.

____Frequencies. Discuss AM radio frequencies and compare with NDB frequencies.

____Advantage of ADF, *not* line of sight. Signals may be picked up when station is over the horizon.

____Disadvantages. Thunderstorms "attract" the needle; Precipitation and storm static may make reception weak or impossible.

____Homing to the NDB. (*Tracking* will not be covered at this point.)

____Always confirm that the proper station has been selected; identify.

____Station passage. The airplane may not go directly over the station but can be a few hundred yards to either side for an acceptable passage.

____Transponder. "IFF Codes" ("Identification, Friend or Foe"—World War II and Korea).

____Theory of operation.

____4096 codes available (only 6, including Emergency, for World War II Navy fighters).

____VFR code is 1200 for *all altitudes* (the student probably has been using the transponder since starting flying, but the equipment should be reviewed at this point).

____Emergency (7700) and loss of communications (7600) codes and when they should or should not be used.

____Practice setting codes while sitting in the airplane with engine shutdown and master switch OFF.

____Communications procedures to radar facility if lost; Confess, Climb, Communicate, Comply, Conserve, and Cool, keeping of (*SPFM*, Chapter 25).

____Frequencies to use if lost. Call FSS, towers, or approach controls, check sectional chart for possibilities within a 100-NM radius.

____Use of radar services for locating the aircraft.

____Frequencies to be used, FSS, tower, or approach control. If unsure of frequencies, call "nearest Flight Service Station on this frequency": 122.0 (at selected stations, Flight Watch), 122.2, 122.3, 122.6, 123.6, Emergency 121.5 MHz. FSS will listen on 122.05, 122.1, 122.15, and 123.6, and the pilot can receive on the affiliate VOR frequencies. Confirm these frequencies. Assure the student/private pilot that if lost or "uncertain" contact should be made on the listed frequencies and let those folks carry the ball from there. They will give location and all help possible, a successful on-airport landing is assured if the fuel situation has not passed the critical point. *Emphasize that the sooner the steps are taken, the better.* The student will probably be asked to switch frequencies to a radar facility for vectoring. The frequency-switching procedure should be well understood so that communications aren't lost. Student pilots should have all these frequencies available on a card (clipboard, wallet, etc.) when on a cross-country flight.

____VHF/DF services. This is a disappearing service and may not be available for some parts of the country. The student/private pilot should contact the FSS, tower, or approach control for any help available.

____ATC light signals. Review in the classroom (*SPFM*, Fig. 21-21).

____Review communications procedures.

    ____Unicom. Frequencies used at different locations.

    ____Multicom. When is this frequency used?

    ____Air-to-air frequencies.

    ____FSS. Position reports, weather data, or local advisory service.

    ____Communications steps flying into a tower-controlled airport.

    ____Communications steps in departing a tower-controlled airport.

Assigned Reading. Student: *SPFM*, Chapters 21 and 26; *AIM* Chapters 2 and 8. Instructor: *FIM*, Chapter 16.

Comments_____

_____

Instructor_____

Date_____ Ground Instruction Time_____

Student Initials_____

UNIT 3 - Introduction to Night Flying

At least one night dual should be completed before the student flies the first triangular solo cross-country (50-NM leg).

■ **Ground Instruction. 1.5 Hours.** Cover the following areas before the first night dual:

____Airplane equipment.

    ____Required equipment (FAR 91.205) and lighting (FAR 91.209) for VFR night flying.

    ____Personal equipment, flashlight, charts, checklist.

    ____Procedures in the event of loss of electrical power.

____Physiology of night flying.

    ____Rods and cones.

____How to effectively scan for other aircraft.

____Visualizing the position and course of another aircraft by its position lights.

____Vertigo.

____Indications of flying into fog or mist (haloes around lights).

____Illusions in flight.

   ____The need to rely more on the flight instruments at night.

   ____Sparse ground lights and disorientation.

   ____Landing errors and illusions. Use of VASI and electronic glide slopes.

   ____Runway up- or down-slopes and narrow and wide runways.

   ____Atmospheric illusions. Rain on windshield, landing in fog.

   ____Don't land on a road or parking lot, use the runway. It has happened!

   ____"Leans."

   ____Acceleration after takeoff. An illusion of nose-up attitude.

____Cockpit lighting. Variable controlling for different night conditions.

____Airport lighting

   ____Airport rotating beacons, civilian, military, and water airports. Rotating beacons may be on in the daytime for Class B, C, D, and E airspace when conditions are below VFR.

   ____Runway edge lights.

   ____VASI. Various types available.

   ____Runway end lighting.

   ____Approach light systems.

   ____Taxiway and other surface lights.

   ____Pilot control of runway lighting systems.

____Discuss short cross-country to nearby town and return.

____Takeoffs and landings.

____Discuss go-arounds at night.

■ **Flight: Dual. 1.0 Hour.**   This first night flight should start in twilight so that a gradual transition to darkness occurs; however, a minimum of 1.0 hour in legal night flight should be accomplished, so that the total flight time could be up to 2 hours in some cases.

____Preflight check of all exterior and interior lighting. Flashlight and other personal equipment within easy reach. Cyalume® stick should also be available.

____Prestart and starting procedures. Navigation (position) lights ON before engaging starter. All other electrical equipment OFF. Other lights (rotating beacon, taxi) ON after start.

____Taxi.

____Run-up (taxi and/or landing lights OFF).

____Takeoff. Using landing light(s).

____Climb.

____Departure pattern.

____Short cross-country, using pilotage (and VOR, and nondirectional beacon [NDB] as available).

____Reenter pattern.

____Shoot at least five takeoffs and landings to a full stop, using landing lights and interior lights.

____Demonstrate go-around procedure.

____Shutdown and secure the airplane.

____Postflight check and security of the airplane.

____Postflight instruction.

    ____Evaluation.

    ____Critique.

    ____Review.

Assigned Reading. _Student: SPFM, Chapters 19 - 25. AIM Chapters 3 and 4. Instructor: FIM, Chapters 13 and 14._

Comments_____

_____

Instructor_____

Date_____ Ground Instruction Time_____ Flight Time_____

Student Initials_____

UNIT 4 - Navigation Groundschool: Preparation for Dual and Solo Cross-Countries

■ **Ground Instruction. 3.0 Hours.** This should be given in two or more sessions.

____Types of navigation.

    ____Pilotage.

    ____Dead (deduced) reckoning.

    ____Celestial navigation, briefly.

    ____Radio navigation.

____Meridians and parallels.

    ____Meridians run north/south and converge at the true North and South Poles.

        ____Prime meridian runs through Greenwich, England.

        ____Lines of longitude run north and south; longitude is measured from the prime meridian east and west.

        ____Time zones/universal coordinated time.

    ____Parallels (lines) run east/west.

        ____Prime parallel runs through the equator.

        ____Latitudes are measured north and south of the equator.

____Degrees of latitude.

    ____60 minutes = 1 degree.

    ____1 minute = 1 nautical mile (NM) or 1.152 statute miles (SM).

____True North. Courses are measured clockwise from true north.

____Magnetic North Pole is in Canada.

____Magnetic variation = difference between Magnetic North and True North. This is not necessarily the *geometric* angle between the two poles since the isogonic lines "meander."

    ____Isogonic lines = lines of equal magnetic variation.

    ____Agonic line.

    ____East is least, West is best. Subtract from the true course or heading for easterly variation; add for westerly.

____Compass deviation.

    ____Swinging or correcting a compass.

____TVMDC.

____Plotter.

    ____Measuring the true course on the mid-meridian.

    ____Mileage scales. Sectional and other charts; nautical and statute miles.

____Wind triangle.

    ____Vector system.

    ____Wind directions (winds aloft) are given in true directions and in knots.

    ____Selecting VFR cruising altitudes.

        ____0°–179° = odd thousands plus 500 feet.

        ____180°–359° = even thousands plus 500 feet.

        ____Selecting the altitude with most-favorable winds.

    ____Draw a True North/South line on a blank sheet of paper for reference.

    ____Use a plotter to measure and lay out the true course line through some point on the north/south line. Draw this line of indefinite length.

    ____Using the nautical mile scale on the plotter, draw the wind vector in the proper direction and length from some point on the course line.

    ____From the head of the wind arrow, swing a line the length of the true airspeed value (use the nautical mile scale on the plotter) until it hits the true course line.

    ____Discuss the meaning of each line in the triangle.

It is suggested that this might be a good break point for this Unit.

____Using the E6B and other computers.

    ____Finding true airspeed.

    ____Checking fuel consumption and speed.

    ____Finding density-altitude.

    ____Finding groundspeed and estimated time of arrival (ETA).

    ____Finding fuel consumption.

    ____Correcting indicated or pressure altitude for temperature.

____Converting the true course to true heading to compass heading and back.

____Sectional chart.

    ____Aeronautical symbols.

    ____Chart coverage and scale.

    ____Topographical symbols.

    ____Radio aids to navigation and airspace information.

    ____Data on obstruction symbols.

    ____List of ATC frequencies.

    ____Prohibited, restricted, warning, and alert areas.

    ____A, B, C, D, E, and G classes of airspace.

____*Aeronautical Information Manual (AIM)*—Basic flight information and ATC procedures.

____*Airport/Facility Directory (A/FD).*

____Review legend and check information for airports on the first dual and solo cross-country flights.

____Notices to Airmen (NOTAMs). L, D, and the FDC publication.

Another suggested break point for Unit 4.

____National Transportation Safety Board Part 830.

____The airplane.

    ____Performance factors for takeoffs and landings.

    ____Systems.

    ____Weight and balance.

        ____Seesaw principle.

        ____Datum moments and CG positions.

        ____Empty weight CG and moment.

        ____Basic empty weight.

        ____Useful load.

        ____Fuel and oil weights.

        ____Effects of adding, subtracting or moving weights.

Use the following as a checklist.

____Airplane papers required on board (A ROW):

    ____Airworthiness certificate (A).

    ____Registration certificate (R).

    ____Operations limitations (O).

    ____Weight and balance (W).

____Emergency procedures.

____Weather information sources.

    ____Hourly reports.

    ____Forecasts.

    ____Weather charts.

    ____Winds aloft.

    ____In-flight advisories.

____Filing flight plans.

    ____Advantages of filing. Procedures for opening and closing a VFR flight plan.

    ____Use of transponder in emergency situations.

    ____Review cockpit resource management (CRM) and aeronautical decision making (ADM) as applied to student and low time private pilots.

Assigned Reading. Student: *SPFM*, Chapters 19-25. *AIM; A/FD*. Instructor: *FIM*, Chapters 13 and 14.

Comments_____

_____

Instructor_____

Date_____ Ground Instruction Time_____ Flight Time_____

Student Initials_____

UNIT 5 - Dual Cross-Country to Two Other Airports, One of Which Is at Least 50 NM from the Airport of Origin

■ **Ground Instruction. 1.5 Hours**

____Sectional chart.

    ____Measure the true course and distances for the three legs and mark the chart. (It is suggested that markers be placed every 10 NM.)

    ____Plan at least one leg on a federal airway.

    ____Track on a VOR or home on NDB on one leg, if ground and airborne equipment is available.

    ____Check alternate airports.

    ____Set up flight log (*SPFM*, Fig. 24-19)

    ____Select enroute check points and mark on chart.

    ____Operational problems associated with different terrain features.

    ____Operations at high density-altitude airports, including leaning of fuel mixture for best power for takeoff.

    ____Estimating visibility while in flight, tips.

    ____Check for restricted, prohibited or special-use airspace on the routes or at the destinations.

    ____Review chart legends, including listings of control tower frequencies and Class B, Class C, and selected radar-approach control frequencies.

____*Airport/Facility Directory.*

    ____Check facilities at destination. Review with student the airports and Fixed Base Operators [FBOs], location of wind indicators, Unicom or approach control/tower frequencies.

    ____Review radio navigation aids available at destination and enroute.

____Flight Service Station. Personal visit or the student calls, with the instructor on the phone extension.

____Current weather enroute and at the two destination airports.

____Area and terminal forecasts (if available for destination airports or if not, get at closest airports). Obtain area forecasts and terminal forecasts for trip time plus 2 hours. Write them down.

____Winds aloft forecast.

____Check NOTAMs, enroute and at destination airports.

____Compute climbs, fuel used, enroute times, and headings. Decide on refueling points.

____File flight plan, agree on activation procedures. Leave plenty of time prior to ETD for student to complete flight log.

■ Flight: Dual. 2.5 Hours

____Checklist for personal equipment, charts, logs, computers, plotters, and a reliable timepiece.

____Preflight check (full fuel and oil).

____Airplane servicing requirements.

____Start.

____Taxi.

____Pretakeoff check.

____Takeoff.

____Fly over the airport enroute, start timing.

____Checkpoints and times.

____Estimating visibility while in flight.

____Alternate airport procedures.

____Track VOR or home on NDB.

____Flight under the hood on part of one or more legs.

____Locating position by VOR cross-bearings.

____Traffic pattern at destination airports.

____Close flight plan(s). If possible always close flight plan after landing at destination (but don't forget).

____Refuel at one or both airports, for student training purposes.

____Return to home airport.

____Postlanding procedures. Postflight check and securing of airplane.

____Fill out logbook for completion of dual cross-country.

Note: The instructor may elect to fly a second dual cross-country to the same airports before sending the student on the solo cross-country.

____Postflight instruction.

    ____Evaluation.

    ____Critique.

    ____Review.

Assigned Reading. Student: *SPFM*, Chapters 24 and 25. *A/FD*, Legends. Instructor: *FIM*, Chapter 14.

Comments_____

_____

Instructor_____

Date_____ Ground Instruction Time_____ Flight Time_____

Student Initials_____

UNIT 6 - Solo Cross-Country

■ **Ground Instruction. 1.0 Hour.** Review first dual cross-country and work with student to plan a solo cross-country to those same airports.

____Review airplane servicing requirements.

    ____Fueling and proper grounding of the airplane.

    ____Oil.

    ____Starting the engine with external power and when *not* to use external power (when the battery is completely dead).

    ____Proper windshield cleaner.

    ____Special requirements for a particular airplane.

____Weather (current and forecast), enroute and destinations.

____Winds aloft. Student chooses altitudes (odd +500, even +500 as applicable) for each leg.

____Student prepares flight log.

____If student lands at an airport (or other point) other than the destination airports he is to call a flight instructor at home base and *not depart* until cleared.

____Review VOR or NDB, communications frequencies, and lost aircraft procedures.

____Alternate airports.

____Student files flight plan with FSS for each separate leg *or* a round robin under instructor agreement and supervision.

____Discuss advantages and disadvantages of filing for the full trip versus filing for each leg.

____Checklist for charts, computers, time piece, and logbook.

____Pilot logbooks and pilot/medical certificates.

____Money or credit card(s) for refueling.

____Review paperwork and sign student pilot certificate and logbook, describing the allowed cross-country.

■ **Flight: Solo. 3.0 Hours**   (4.0 hours total elapsed time; allow more time than for the dual cross-country).

____Preflight check (*full* fuel and oil).

____Activate flight plan after takeoff.

____Student flies to first destination airport, closes flight plan, and gets logbook signed.

____Refuels at first airport, if planned. Files flight plan for second airport, if not on a round-robin flight plan.

____Departs for second airport without undue delay.

____Opens flight plan.

____Closes flight plan, refuels at second airport (if planned or necessary) and gets logbook signed.

____Files flight plan for home airport as required.

____Departs for home airport.

____Opens flight plan.

____Closes flight plan after arrival.

____Postflight procedures.

____Instructor and student review and debrief the flight.

    ____Evaluation.

    ____Critique.

    ____Review.

Assigned Reading. Student: *SPFM*, Chapters 21 - 25. Instructor: *FIM*, Chapters 13 and 14.

Comments_____

Instructor_____

Date_____ Ground Instruction Time_____ Flight Time_____

Student Initials_____

UNIT 7 - STAGE Check after the First Solo Cross-Country

FAR Part 141 requires that a flight test be given at the completion of the first solo flight, at the completion of the first solo cross-country flight, and at the conclusion of the course.

For this postsolo cross-country stage check, the check pilot will have the student pilot plan and fly a short cross-country, utilizing current weather and forecasts, including winds aloft information, NOTAMs, *Airport/Facility Directory*, and latest sectional chart(s). The check pilot may, during the flight, require that the student divert to another airport to cope with various simulated inflight problems and emergencies, such as engine overheating, partial power failure, electrical failure, carburetor ice, and unforecast deteriorating weather. The flight will return to home base without landing at other airports unless the check pilot feels it is necessary for a particular student. The use of VOR and/or NDB will be checked, as applicable.

■ Flight: Dual. 1.5 Hours

____Weather planning.

    ____Current weather enroute and at destination.

    ____Forecast weather enroute, destination, and home airport for the proposed flight period plus 2 hours.

    ____Winds aloft forecasts for the terminal(s) nearest to the route.

____Preflight navigation planning.

    ____Drawing course line(s).

    ____Using a plotter.

    ____Computer.

    ____Expected fuel consumption.

    ____ETAs to checkpoints, destination, and home base.

____Preflight check.

____Cockpit management.

____Pretakeoff check.

____Procedures to climb and get established on course.

____Follow the preplanned course by use of landmarks; verify position within 3 NM.

____Altitude control (±100 feet).

____Heading control (±10°).

____Enroute checkpoints and destination ±5 minutes.

____Landmark identification.

____On-course accuracy.

____Correct and record differences between preflight fuel, groundspeed, and heading.

____Diversion to an alternate airport.

____Use of VOR or NDB for bearing, tracking or homing to get to an alternate airport.

____Emergencies. Choose one of the following to be simulated:

    ____Complete power failure.

    ____Partial power failure.

    ____Electrical failure.

    ____Overheating engine.

    ____Weather deterioration.

    ____Procedures, including communications, when lost.

____Return to home airport.

    ____Checklist.

    ____Communications.

    ____Traffic pattern.

    ____Landing.

    ____Postlanding procedures. This is necessary after *every* flight as was noted in STAGE 1.

        ____Checklist.

        ____Postflight inspection.

        ____Security of the airplane.

    ____Postflight instruction.

        ____Evaluation.

        ____Critique.

        ____Review.

Assigned Reading.   Student: *SPFM,* Chapter 15. Instructor: *FIM,* Chapters 12 and 24._____

Comments_____

_____

_____

Instructor_____

Date_____ Ground Instruction Time_____ Flight Time_____

Student Initials_____

UNIT 8 - Emergency Instrument Flying

■ Ground Instruction. 1.0 Hour.

____Review Four Fundamentals.

____Instrument scan.

____Six Basic Maneuvers. Brief on using both full and emergency panels.

   ____Straight and level flight, ±10° of heading.

   ____Left and right 180° turns, ±10° on rollout.

   ____Shallow climbing turns to a predetermined altitude and heading.

   ____Shallow descending turns to a predetermined altitude and heading.

   ____Recovery from the start of a power-on spiral.

   ____Recovery from the approach to a climbing stall.

____Tracking to a VOR, hooded at a constant altitude.

____Homing to an NDB, hooded at a constant altitude.

____Radar vectors, hooded to an ASR approach. The instructor may simulate ATC.

   Aviate, Navigate, Communicate!

Flight: Dual 1.5 Hours.   This Unit may use a flight simulator and authorized instructor if available (FARs 61.4 and 61.109).

____Student puts on hood during climbout and flies specified headings (full panel).

____Climb to safe altitude and fly to practice area, vectored by instructor.

____Six Basic Maneuvers, hooded full panel, followed by emergency panel practice.

____Tracking to a nearby VOR, full panel.

____Homing to an NDB (if aircraft and ground equipment is available), full panel.

____Use of communications while flying the airplane, hooded and full panel.

____"Radar" vector by instructor to home airport.

   ____Evaluation.

   ____Critique.

   ____Review.

Assignment for the next Unit: Plan a triangular solo cross-country to two airports not previously visited.

Assigned Reading. Student: *SPFM*, Chapters 22 - 25. Instructor: *FIM*, Chapters 13 - 14.

Comments_____

_____

Instructor_____

Date_____ Ground Instruction Time_____ Flight Time_____

Student Initials_____

UNIT 9 - Solo Triangular Cross-Country
(A solo cross-country to two airports not previously visited)

■ Ground Instruction. Preflight Briefing. 1.0 Hour.

____Airplane papers. Review fuel and oil requirements.

____Review student's sectional chart planning and checkpoints on log, refueling procedures, and alternate courses of action.

____Check Airport/Facility Directory for the destination airports plus alternate airports and communications procedures.

____Weather check (current and forecasts) by student, with instructor review.

____Winds aloft information checked, then enroute altitudes chosen.

____NOTAMs, enroute and at airports.

____Student completes the flight log.

____Check flight log and ETAs.

____Student files flight plan to first destination.

____Instructor signs logbook for this specific flight route and date. (Double check student pilot certificate for the cross-country sign-off.)

____Review collision avoidance and wake turbulence avoidance procedures.

____Check that student has money and/or credit cards for fuel.

■ Flight: Solo. 3.0 Hours (4.5 hours elapsed time for the trip). The student completes the following:

____Preflight.

____Departure.

____Navigation to first destination.

____Traffic pattern and landing.

___Close flight plan.

___Logbook signed.

___Fuel as necessary.

___Check second destination weather and forecast.

___File flight plan to second destination.

___Depart and open flight plan, or open flight plan just before departure. Note that Flight Following Service on student cross-countries is discouraged.

___Navigate to second destination, using pilotage (and radio navigation as available).

___Traffic pattern and landing.

___Close flight plan.

___Logbook signed.

___Fuel as necessary.

___Review route to home airport.

___File flight plan: check enroute and home airport weather.

___Depart and open flight plan.

___Navigation to home airport using pilotage and radio navigation.

___Land at home airport and close flight plan.

___Instructor reviews the flight and, after the airplane is fueled, compare actual fuel used with predicted fuel consumption.

___Instructor signs logbook for arrival at home.

　　___Evaluation.

　　___Critique.

　　___Review.

Assigned Reading.　Student: *SPFM*, Chapter 22 - 26, *AIM*, "Lighting". Student prepares a flight log and reviews the sectional and *A/FD* for a dual night cross-country flight to an airport 50+ NM distant. Instructor: *FIM*, Chapter 16.

Comments_____

_____

Instructor_____

Date_____Ground Instruction Time_____Flight Time_____

Student Initials_____

UNIT 10 - Dual Night Cross-Country and Local Pattern Work (Initiate actual flight in full darkness)

■ Ground Instruction. 1.0 Hour

____Preflight briefing for a dual night cross-country to an airport 50+ NM miles away to shoot take-offs and landings (full stops); return to home airport for more landings. If possible, fly to an airport with pilot-controlled lighting during one of the night flights.

____Check weather, current and forecast, for destination and home airports.

____Check charts and *Airport/Facility Directories.*

____Review student's flight log previously prepared for this flight (except for winds aloft information).

____Assure that personal equipment (flashlight, timepiece, etc.) are available.

____File round-robin flight plan, including time for takeoffs and landings at the other airport and at home, as desired.

■ Flight: Dual 2.5 Hours

____Preflight check.

  ____Airplane's interior and exterior lighting.

  ____Discuss problems of preflight checking at night with a flashlight versus full daylight pre-flighting.

____Prestarting procedures. Master switch and interior lights ON. Flashlight is needed for hard-to-see items, such as fuel selectors and other controls.

____Navigation (position) lights ON, indicating that pilot is onboard and ready to start.

____After start, all necessary exterior lights ON.

____Taxi. Taxi/landing lights ON. Use extreme caution.

____Run-up. Taxi/landing lights OFF. Use flashlight to check controls (ailerons, elevator, rudder) for proper movement and for interior hard-to-see switches or controls.

____Takeoff. Check area for other airplanes. Landing light(s) ON before taking the runway.

____Open the flight plan.

____Depart for destination airport, noting relative position of nearby towns while enroute.

____Contact approach control, tower, Unicom or broadcast in the blind on the CTAF as applicable.

____Enter traffic, using proper communications and utilizing collision avoidance procedures.

____Shoot at least three full-stop landings, depending on the size of the airport, whether it is tower-controlled, or other factors.

____Return to home airport.

____Shoot first two patterns (full stop) with full cockpit lighting and landing/taxi light.

____Shoot at least two patterns (full stop) without landing/taxi light. Use landing/taxi light for taxi back.

____Two patterns (full stop) with landing/taxi lights but interior lights OFF. (Use exterior references and red lensed flashlights.)

____Two patterns with landing/taxi lights *and* interior lights OFF. Keep navigation/position lights and rotating beacon ON.

____One pattern (full stop) with full interior lights restored.

____Go-around by student, followed by full-stop landing and securing of the airplane.

____Close flight plan.

____Instructor debriefs after flight and fills out logbook.

       ____Evaluation.

       ____Critique.

       ____Review.

Assigned Reading.   Student: *SPFM*, Chapters 10 and 13. Instructor: *FIM*, - Chapters 13, 14, and 25.

Comments_____

_____

Instructor_____

Date_____Ground Instruction Time_____Flight Time_____

Student Initials_____

UNIT 11 - Solo in the Practice Area and Pattern V

This flight is to assure that the student maintains proficiency in the maneuvers previously discussed and flown. It provides a "break" from the night and cross-country work and maintains student confidence in slow flight, stalls, wind drift correction maneuvers and normal and special takeoffs and landings. The briefing is to review the maneuvers and resolve any questions; it may be longer than earlier briefings for practice area and pattern work.

■ Ground Instruction. 1.0 Hour

____Pattern and departure.

____Slow flight, clean.

____Takeoff and departure stalls.

____Approach-to-landing stalls.

____360° and 720° power turns.

____Descent to wind drift maneuver altitude (600 to 1000 feet AGL).

____Carburetor heat, as needed.

____Rectangular course.

____S-turns across the road.

____Turns around a point.

____Return to airport.

    ____Checklist.

    ____Communications.

    ____Pattern entry.

____Normal takeoffs and landings, touch-and-go or full-stop.

____Crosswind takeoffs and landings, as applicable.

____Short-field takeoffs and landings, full stop.

____Soft-field takeoffs and landings, full stop.

____Postlanding procedures.

____Postlanding checklist.

____Securing the airplane.

■ Flight: Solo. 1.0 Hour

____Preflight and pretakeoff checklists.

____Depart traffic pattern and climb to practice area.

____360° and 720° power turns.

____Slow flight, clean.

____Takeoff and departure stalls.

____Approach-to-landing stalls.

____Descend to the altitude required for wind drift correction maneuvers, airspeed ±10 knots of best glide speed. Use carburetor heat as directed by the *POH*.

____Wind drift correction maneuvers (time may not allow practice of all three listed), airspeed ±10 knots, altitude ±100 feet.

    ____Rectangular course.

    ____S-turns.

    ____Turns around a point.

____Return to pattern.

____Normal or crosswind takeoffs and landings, as applicable.

____Short-field takeoffs and landings, full stop.

____Soft-field takeoffs and landings, full stop.

___Postlanding procedures.

    ___Postlanding checklists.

    ___Securing the airplane.

___Postflight review by instructor.

Assigned Reading.___ Student: *SPFM,* Chapters 21 and 24. *AIM* Chapter 4 (parts as assigned by instructor. Instructor: *FIM,* Chapters 13 and 14.

Comments_____

_____

_____

Instructor_____Cert. No._____Exp._____

Date_____Ground Instruction Time_____Flight Time_____

Student Initials_____

UNIT 12 - Dual and Solo Flight to a Tower-Controlled Airport (VFR-Day)

The outline of this unit assumes that the home airport does not have an approach control/tower and that this is the student's first experience with a controlled airport. In this case the instructor may elect to shoot at least three dual takeoffs and landings to a full stop with taxi-back at the big airport so that the student feels comfortable with operations and communications. It's suggested that time be set aside during this lesson for a visit to the approach control/tower facilities to allow the student to meet the personnel and to feel more comfortable in operations at that airport. A student pilot who has been operating out of a controlled airport will have already met the requirements of FAR 61.109, which requires three solo takeoffs and landings to a full stop at an airport with an operating control tower.

■ Ground Instruction. 1.0 Hour

___Steps when approaching and landing at a controlled airport.

    ___ATIS (Airport Terminal Information Service).

    ___Approach control. Assigns a discrete transponder code on initial contact.

    ___Tower.

    ___Ground Control.

___Procedures during takeoff and landing operations in the pattern.

    ___Ground control; contact when directed by the tower.

    ___Tower.

____Steps when departing the controlled airport.

    ____ATIS.

    ____Clearance delivery. Indicate that Information Delta, etc., has been received.

        ____Heading and altitude are given.

        ____Departure control frequency (and discrete transponder code) given here. Set the code but keep the transponder on standby.

    ____Ground control.

    ____Tower.

    ____Transponder ON at start of takeoff.

    ____Contact departure control when directed by tower on climb-out.

    ____Transponder, 1200 after release by departure control.

____Wake turbulence avoidance.

____Light signals - Review. Assure that light signals card is in the aircraft.

____Describe layout of taxiways and location of FBOs at the large airport.

    ____Have diagram of the airport, if available.

____Confirm available sectional charts (IFR approach charts also, if available) are current.

____Check log of the trip.

____Check destination airport information in the *Airport/Facility Directory.* Discuss any special requirements for that particular airport.

____Check *AIM* for airport markings and signs. Take copy of *AIM* on flight for reference to "unusual" markings.

■ Flight: Dual. 2.0 Hours; Solo. 1.0 Hour

____Student conducts preflight check, pretakeoff check, takeoff, and climb.

____Navigate to tower-controlled airport.

____Student listens to ATIS, confirms information and contacts approach control at proper time and position.

____Student selects discrete transponder code as given by approach control.

____Student communicates, lands, taxies to destination on airport, and secures airplane, under supervision.

____Fuel as necessary.

____Tower visit.

____Student returns to airplane and shoots at least three solo takeoffs and landings to a full stop. Instructor may elect to stay in tower cab during the takeoffs and landings.

____Returns to FBO to pick up instructor *before* departing for home airport. This is very important!

____Navigates directly to home airport. It's suggested that no side trips or alternate routes be taken since the student is probably fatigued by this "busy" session.

____Postlanding procedures.

____Review with instructor all parts of trip after landing.

    ____Evaluation.

    ____Critique.

    ____Review.

Assigned Reading.   Student: *SPFM*, Chapters 22 and 23; read selected portions of the latest copies of *Aviation Weather Services (AWS)* AC 00-45 and *Aviation Weather (AW)* AC 00-6

Comments_____

_____

Instructor_____

Date_____Ground Instruction Time_____Flight Time_____

Student Initials_____

UNIT 13 - Weather Theory and Weather Services

■ **Meteorology.**   This Unit will consist of ground instruction on basic meteorology and weather services available. Particular attention will be paid to hazards to flight and how to recognize them both on the ground and in flight. These subjects should be covered in two *or more* ground school sessions, but they should be completed before the long cross-country.

■ **Ground Instruction. 6.0 Hours**

____Basic meteorology *Aviation Weather (AW)* AC 00-6.

    ____Heat and circulation.

    ____Coriolus effect in Northern and Southern hemispheres.

    ____Moisture: temperature and dewpoint effects on the atmosphere's ability to produce or hold moisture.

    ____Relative humidity.

    ____Lapse rates: normal, dry, and wet.

    ____Clouds, families.

        ____How clouds are formed.

        ____Clouds, cumulus and stratus forms.

____Fog: advection, radiation, upslope, precipitation and ice fog.

____Precipitation: rain, hail, sleet, and snow.

____Fronts: Warm, cold, occluded, and stationary; the clouds and visibilities (and precipitation) expected with each.

    ____Cold front: type of clouds and weather, turbulence, area covered and precipitation. Cross-section of the front.

    ____Warm front: expected clouds and weather and extent of area covered. Freezing rain in winter may occur. Cross-section.

    ____Occluded front - May contain weather of both warm and cold fronts.

    ____Frontogenesis and frontolysis.

Suggested break point.

____Hazards to flight.

    ____Thunderstorms.

        ____How thunderstorms are formed, the three requirements are instability, lifting force, and moisture.

        ____Cumulus clouds. Mature and dissipating stage, updrafts, downdrafts, and precipitation. Mature stage is the most hazardous.

        ____Dissipating stage, the anvil head. *Do not* fly under the anvil head; stay *at least* 5 miles away (10 miles is even better).

    ____Hail. How it is formed.

    ____Icing. Clear, rime, and mixed.

    ____Freezing rain: When it may occur.

    ____Lightning.

    ____Turbulence.

    ____Windshear (loss of airspeed).

    ____Microbursts.

    ____Fog: A 4° F temperature and dewpoint spread could indicate that fog formation may start.

    ____Major hazards of fog to the student and private pilot.

    ____Microbursts. How they occur.

    ____Turbulence in clear air conditions (*SPFM*, Chapter 23).

Break point.

____*Aviation Weather Services (AWS)* AC 00-45.

    ____Review a typical hourly routine weather report (METAR).

    ____TAFs (Terminal Aerodrome Forecasts).

    ____Area forecasts.

____Inflight advisories.

   ____EFAS (enroute flight advisory service).

   ____SIGMETs (*significant meteorological* information).

   ____AIRMETs (*meteorological* phenomena that are potentially hazardous to *air*craft).

   ____Convective SIGMETs.

   ____TWEBs (transcribed weather broadcasts).

   ____AWOS (automated weather observation system).

   ____ASOS (automated surface observation system).

____PIREPs (pilot reports).

____Wind information.

   ____Winds aloft forecasts.

   ____Winds aloft charts.

____Other services.

   ____TIBS (telephone information briefing service).

   ____HIWAS (hazardous inflight weather advisory service).

   ____Flight watch, 122.0 MHz.

Break point.

____Weather charts.

   ____Surface analysis chart.

   ____Weather depiction chart.

   ____Radar summary chart.

   ____Weather prognostic chart.

____Evaluation.

____Critique.

____Review.

Assigned Reading. Student: *SPFM,* Chapters 21-25. FAR 61 and 91. Instructor: *FIM,* Chapters 13, 14 and 25.

Comments_____

_____

Instructor_____

Date_____Ground Instruction Time_____

Student Initials_____

UNIT 14 - Solo Long Cross-Country Flight

■ **Ground Instruction. 1.0 Hour.** This flight is suggested for a cross-country of at least 300 NM with landings at a minimum of three airports, one of which is at least 100 NM from the original departure point. These were the minimum distances required in the pre-1997 revision of FAR 61.109. The current requirements are a solo cross-country flight of at least 150 NM total distance with full-stop landings at a minimum of three points, with one leg of the flight being a straight-line distance of at least 50 NM between takeoff and landing locations. The minimum solo cross-country flight time is now 5 hours, not 10 hours, as previously required. The instructor may use either time span, depending on the student, flight time required, and conditions of training, but the instructor will brief the student and ensure that the flight will be completed well before dark.

____Review sectional chart(s) and initial navigation log with student. Check for restricted and prohibited areas.

____Flight instructor and student together complete weather briefing with current weather and forecasts (terminal and area) and check NOTAMs for enroute and destinations.

____Instructor and student check winds aloft and times enroute to complete the navigation log.

____Assure that the airplane papers are in order and that a credit card or cash is available for fueling at prechosen destination airport(s). *Suggested:* refueling at each airport recommended.

____Recheck student pilot certificate for initial endorsement; sign logbook for this route and date.

____Confirm that the student logbook has latest 90-day endorsement.

____Student pilot has phone number(s) where CFI can be located during the time period of the trip.

____Student files flight plan for the first leg and will file for the other two legs in order.

____Flight instructor will be available at airport to debrief the student upon return.

■ **Flight: Solo. 4.0 Hours**

____Student flies the cross-country.

____Review the cross-country flight with student upon return.

Comments_____

_____

_____

Instructor_____

Date_____Ground Instruction Time_____

Student Initials_____

UNIT 15 - STAGE 3 Written Test

■ **Ground Instruction Credit. 1.0 Hour.**   A written test made up by the school or instructor will be given on the cross-country phase to check that the student fully understands the requirements for a safe VFR cross-country. The test should cover the following areas as a minimum:

____Weather services available.

____Preflight weather briefing.

____FAA flight plans.

____Sectional chart.

____Navigation.

    ____Pilotage.

    ____Radio navigation.

____Communications with the FSS and/or tower.

____Emergency procedures.

    ____Weather encounters.

    ____Carburetor ice.

    ____Diversion to other airports.

    ____Electrical failure or fire.

    ____Rough-running engine.

    ____Lost procedures.

        ____Communications.

        ____Precautionary off-airport landing.

            ____Dragging the area.

            ____Postlanding procedures.

Grade_____

Date_____Instructor Signature_____

Cert. No._____Exp._____

STAGE 4

PREPARATION FOR THE KNOWLEDGE TEST

PRIVATE PILOT, ASEL

■ **Ground Instruction. 9.0 Hours**

This ground instruction preparation for the Knowledge Test session may be broken up into eight 1-hour or four 2-hour sessions or reviews, plus one 1-hour session for summary and questions. Break up the subject matter and arrange it to fit the student's and your schedule: because of this, you may not exactly match the order listed here. Depending on the student, it may be recommended that the Knowledge Test be taken anytime during the postsolo period, but it is suggested that at least the first solo cross-country be completed before taking the test. It's hoped that this STAGE will not be just for preparation for the Knowledge Test but will present information to be used throughout the pilot's career. The questions, answers, explanations and references for the Knowledge Test questions (airplane) are in Chapter 27 of *The Student Pilot's Flight Manual, Seventh Edition.* The following topics should be reviewed:

____Federal Aviation Regulations 1, 61, 71, and 91 and NTSB 830, as applicable to student and private pilots.

____Categories and classes, as indicated for certification of pilots *and* aircraft.

____Definition of night time.

____V speeds. V_A, V_{FE}, V_{LE}, V_Y, V_X.

____Preventive maintenance.

____Pilot documents required in flight and requirements for presentation.

____Duration of medical certificates.

____Requirements for flying a high-performance airplane.

____Biennial flight reviews.

____Recent flight experience. Day, night, and in tailwheel airplanes.

____Change of address.

____Glider towing.

____Private pilot general limitations.

____Federal airways dimensions.

____Final authority for aircraft operations.

____Inflight emergencies and pilot authority. When is a report of a deviation required?

____Aircraft limitations information sources.

____Dropping objects from an aircraft.

____Consumption of alcohol by pilot or passenger.

____Preflight action required of the pilot.

____Seatbelts and shoulder harnesses.

____Formation flight.

____Aircraft rights-of-way.

____Maximum airspeeds in, or underlying, Class B airspace.

____Minimum safe altitudes.

____Altimeter settings.

____ATC clearance deviations.

____Tower light signals.

____Visual approach slope indicator (VASI) systems.

____Non-tower airport traffic patterns.

____Fuel requirements, VFR day and night.

____VFR cloud clearances and visibility requirements for operations below and above 10,000 feet MSL.

____VFR cruising altitudes.

____Papers required to be onboard the aircraft.

____Emergency locator transmitter (ELT) battery requirements.

____Instrument and equipment requirements. VFR day and night.

____Inoperative equipment.

____Aircraft position lights. When required to be ON.

____Supplemental oxygen requirements; altitudes and times.

____Transponders (and encoding altimeters). When are they required?

____Aerobatic flight, airspace and altitudes; parachute repacking requirements.

____Restricted and experimental aircraft limitations.

____Aircraft inspections and repairs. Annual and 100-hour inspections.

____Airworthiness directives.

____Aircraft accidents and incidents.

Suggested break point for the review.

____Aerodynamics and airplane systems. *SPFM*, Chapters 2, 3, 4, 5, 7, 23, and 25 and Appendix D.

    ____Four Forces.

____Angles of attack and incidence.

____Four Forces in straight and level flight.

____Wing frost and its effect on takeoff performance.

____Torque and P-factor.

____Airplane stability.

____Rudder.

____Load factors in turns.

____Wing flaps, operations and purpose.

____Engine operations, temperatures and mixture effects.

____Ignition system.

____Running a tank dry in flight.

____Carburetors and mixture controls.

____Carburetor icing. Conditions favorable for, and indications of, carburetor icing.

____Carburetor heat. Effect on fuel/air mixture.

____Minimum fuel grades for an engine.

____Detonation.

____Fuel tanks. Moisture condensation.

____Oil system.

____Density-altitudes and propeller efficiency.

____Pitot-static system: Airspeed, altimeter, and vertical speed indicator.

____Altimeter and various altitudes.

____Setting the altimeter.

____Altitude effects on indicated airspeed at stall.

____Airspeed markings. White Arc, Green Arc, Yellow Arc, and Red Line.

____Maneuvering airspeed. V_A, not marked on airspeed indicator.

Suggested break point.

____Aircraft Instruments. *SPFM*, Chapters 3, 15, 19, and 25.

____Determining flight conditions from illustrations of instruments.

____Magnetic compass.

____Deviation.

____Compass reactions to turns.

____Compass reactions to acceleration or deceleration.

____When is the magnetic compass accurate?

____Gyro instruments: types and theory of operations.

____Airplane operations, performance and stability: *SPFM*, Chapters 2, 3, 6, 9, 12, 17, 19, and 23.

    ____CG position and effects on longitudinal stability.

    ____Density-altitude effects on takeoff, climb, and landing.

    ____Temperature, relative humidity, and pressure-altitude effects on performance.

    ____Finding the effects of temperature on density-altitude from performance charts.

    ____Correcting from indicated to pressure altitude by computer.

    ____Correcting from indicated to pressure altitude using the chart.

    ____What force makes an airplane turn?

    ____Flight control positions for taxiing in varying wind directions.

    ____What causes a spin?

    ____Stall condition of wings in a developed spin.

    ____Stall angle of attack.

    ____Ground effect, performance and stability.

    ____Load factor effect on stall speed.

    ____Angle of attack, what is it?

Suggested break point.

____Weather. *SPFM*, Chapter 22; *AW* AC 00-6; and *AWS* AC 00-45.

    ____Physical process of weather.

    ____Earth pressure variations.

    ____Temperature inversion.

    ____Standard temperature and pressure.

    ____Altitudes. True, absolute, density, pressure, and indicated.

    ____Altimeter. Flying from high to low pressure and vice versa.

    ____Altimeter. Temperature effects.

    ____Coriolis and friction effects on wind.

    ____Clouds and moisture.

    ____Frost formation and its effect on flight.

    ____Dew.

    ____Atmospheric stability.

    ____Cloud formation associated with stable or unstable air masses.

    ____Finding cumulus cloud bases. (Temperature minus dewpoint in degrees C, divided by 2.2 or 2.5, depending on source.)

    ____Cloud families.

    ____Lenticular clouds.

____Fronts and their characteristics.

____Windshear.

____Aircraft structural icing.

____Requirements for formation of cumulonimbus clouds.

____Thunderstorm stages and their characteristics.

____Squall lines.

____Flight in turbulence.

____Fog types.

Suggested break point.

____Weather information and briefings available.

    ____Transcribed weather broadcasts.

    ____Types of weather briefings.

    ____METARs, as applicable.

    ____PIREPs.

    ____TAFs, as applicable.

    ____Area forecasts.

    ____SIGMETs, AIRMETs, and Convective SIGMETs.

    ____Winds aloft forecasts.

    ____Weather depiction charts.

    ____Surface analysis charts.

    ____Significant weather prognostic charts.

    ____Radar summary chart.

    ____Steps in getting a weather briefing.

Suggested break point.

____Sectional chart: *SPFM*, Chapters 19, 20, 21, 24, 25, 27, and chart in back of book.

    ____Wildlife refuges.

    ____Airspace and cloud clearance requirements.

    ____Recognizing airspaces by color and symbols.

    ____Airport legends for tower and non-tower airports.

    ____Terrain elevations. Colors and symbols.

    ____Services available at various airports on the chart.

    ____Obstacles and safe obstacle clearances.

    ____Recognizing various navigation aids by the symbols.

Suggested break point.

____Engine operations. *SPFM,* Chapters 4, 5, 7, 23, and Appendix D.

    ____Engine cooling. Mixture and airspeed effects.

    ____Throttle, mixture, and prop control for a constant-speed propeller.

        ____Constant-speed propeller. Sequence of control usage for adding and reducing power.

    ____Preflight inspections.

    ____Starting. Handpropping or using the starter.

        ____Starting in cold weather.

        ____Using an external power source.

    ____Special preflight checks after aircraft storage.

____Weight and balance: *SPFM,* Chapter 23.

    ____Fuel and oil weight.

    ____Empty weight, standard and basic.

    ____Useful load.

    ____Computing aircraft CG positions by using tables and charts.

    ____Computing moments when aircraft is empty and loaded.

    ____Calculating weight and CG positions to determine the category, Normal or Utility.

    ____Effects of shuffling passengers and baggage around.

    ____Estimating weight and balance for landing.

____Cruise control and fuel consumption. *SPFM,* Chapter 23.

    ____Use of power setting charts.

    ____Determining fuel consumption.

____Takeoff and landing performance. *SPFM,* Chapter 17.

    ____Use of tabular and graphical takeoff and landing charts.

    ____Crosswind component graph use. Calculating headwind and crosswind components.

Suggested break point.

____Safety in flight. *SPFM,* Chapters 10, 13, and 25.

    ____VHF/DF location procedures.

    ____Clearing the area before starting a maneuver.

    ____Power failure after takeoff.

____Aircraft and airport lighting. *SPFM,* Chapter 26, *AIM.*

    ____Scanning at night.

    ____Determining another aircraft's path by its position lights.

____VFR approaches at night.

____Precision approach path indicator.

____VASI. Two- and three-light types and indications.

____Pulsating approach slope indicators.

____Runway high-intensity lights.

____Pilot-controlled lighting.

____Airport rotating beacons.

    ____Civilian.

    ____Military.

    ____Use during daylight hours.

____Airport markings: *SPFM,* Chapter 25, *AIM.*

____Precision runways.

____Nonprecision runways.

____Closed runways.

____Displaced thresholds.

____Taxi lines and taxi-holding-position markings.

____Chevrons (runway unusable for landing, takeoff, or taxiing).

Suggested break point.

____Class B, C, D, E, G, and special use airspace: *SPFM,* Chapter 21, FARs, and *AIM.*

____Special VFR requirements.

____Class B. Minimum pilot and equipment requirements.

____Class C. Vertical and horizontal limits.

____Class C. Satellite airport operations.

____Class D. Lateral dimensions.

____Class D. Non-tower satellite airports.

____Airport Advisory Area Operations (Flight Service Stations).

____Automatic Terminal Information Service (ATIS).

____Alert areas.

____Military operations area (MOA).

____Radar advisories about other traffic and where to look.

____Radar traffic information service.

____Transponder use and codes.

____Airport operations: *SPFM,* Chapters 6, 21, 24, and 25 and *AIM.*

____Communications failure. Procedures when landing at a controlled airport.

____Traffic patterns.

    ____Wind indicators.

    ____Segmented circles.

____Ground control procedures after landing and during taxi.

____Special VFR. Who to contact?

____Flight Plans. *SPFM*, Chapters 24 and 25; and *AIM*.

____Filing.

____Estimated time enroute (ETE).

____Closing the flight plan. With what FAA facility?

____Extending a flight plan – the procedure.

Suggested break point.

____Emergency locator transmitter (ELT): *SPFM*, Chapter 21 and FARs in back of book.

____Transmitting frequencies.

____Battery replacement.

____Testing the ELT. Time frame.

____Monitoring the ELT on 121.5 MHz.

____Wake turbulence and wing tip vortices. *SPFM*, Chapter 13 and Appendix A.

____Creation.

____Actions.

____Worst wind and atmospheric conditions.

____Departing or landing behind a large aircraft.

____Flight Physiology: *SPFM*, Appendix C.

____Carbon monoxide symptoms and effects.

____Scanning for other aircraft.

____Collision course detection.

____Hypoxia.

____Hyperventilation.

____Spatial disorientation.

____Night vision adaptation.

____FAA advisory circulars.

____General numbers for Airmen(60), Airspace(70) and General Operations(90).

____How to obtain them (U.S. G.P.O.)

■ **The Knowledge Test.** The instructor or school should make up a Knowledge Test of 60 sample questions from *The Student Pilot's Flight Manual, Seventh Edition*, Chapter 27, assuring that each area reviewed in this STAGE is covered. Allow 2 hours for this test. A minimum grade of 80 should be required to pass; the missed answers should be reviewed and cleared before endorsing the student to take the FAA Knowledge Test after which the instructor will arrange an appointment with a nearby Computer Testing Designee (telephone numbers are on page 27-6 of *The Student Pilot's Flight Manual, Eighth Edition*) and sign the endorsement for taking the FAA Knowledge Test.

■ **Endorsement for Aeronautical Knowledge: FARs 61.35 and 61.105**

    I certify that I have given Mr/Ms _____the ground instruction required by FARs 61.105 through (5).

    Date _____

    Instructor signature_____

CFI No. _____ Exp._____

STAGE 5

PRIVATE PILOT (ASEL) PRACTICAL (FLIGHT) TEST

UNIT 1 - REVIEW AND INTRODUCTION TO THE PRACTICAL TEST

At least 3 hours of flight instruction (in preparation) are required within 60 days of the practical test.

■ **Ground Instruction. 3.0 Hours.** Review *SPFM*, Chapter 28, and lay out the program for preparation for the practical test, clearing up any questionable areas so that no bad habits have been "set." This review may be done in two or more sessions and should cover the oral portion of the practical test in particular.

■ **Flight: Dual. 2.5 Hours.** This Unit will be an introduction to a portion of the flight section of the Practical Test Standards and the instructor will spend as much time as necessary on each maneuver/flight requirement to assure exceeding the standards set out by the PTS. All of the subjects are covered in detail in *SPFM*, Chapter 28.

____Preflight check. Use of the manufacturer's checklist is required.

____Cockpit management.

____Start. Review cold and hot starting and safety procedures. Use of an external power source.

____Taxi.

____Pretakeoff check. Use checklist.

____Radio communications and ATC light signals.

____Airport and runway markings and lighting.

____Pattern work.

    ____Normal and crosswind takeoff and climb.

    ____Normal and crosswind approach and landing.

    ____Soft-field takeoff and climb.

       ____Airspeed control after takeoff.

    ____Soft-field approach and landing.

       ____Airspeed control.

    ____Short-field takeoff and climb.

       ____Airspeed control after takeoff.

____Short-field approach and landing.

    ____Airspeed control.

____No-flaps approach and landing.

____Forward slips to landing.

    ____Airspeed control.

____Pattern departure.

____Performance and ground reference maneuvers.

    ____Steep turns.

    ____High-altitude emergency.

    ____Rectangular course.

    ____Turns around a point.

    ____Low-altitude emergency.

____Return to traffic pattern.

    ____Radio communications. Unicom, Multicom, or ATC.

    ____Prelanding checklist.

    ____Pattern entry.

    ____Collision avoidance procedures.

    ____Downwind.

    ____Base.

    ____Final.

    ____Touchdown and rollout.

____Postlanding and shutdown procedures.

____Postflight inspection.

____Parking and securing.

____All takeoffs. Directional control.

____All landings. Touchdown and directional control.

____Evaluation.

____Critique.

____Review.

Assigned Reading. Student: *SPFM*, Chapter 28; Practical Test Standards FAA 8081-14, most current. Instructor: *FIM*, Chapter 15.

Comments_____
_____

Instructor_____

Date_____Ground Instruction Time_____Flight Time_____

Student Initials_____

UNIT 2 - Student Practice

■ **Flight. Solo. 1.0 Hour.** The student should practice solo material and maneuvers covered in Unit 1 with a discussion and debriefing by the instructor.

____Evaluation.

____Critique.

____Review.

Assigned Reading.___Student: *SPFM*, Chapters 10, 12, 14, and 28; Practical Test Standards._____

Comments_____
_____

Instructor_____

Date_____Ground Instruction Time_____Flight Time_____

Student Initials_____

UNIT 3 - Student Practice

■ **Flight: Solo. 1.0 Hour.** The student should practice the Practical Test maneuvers assigned by the instructor, based on performance of Units 1 and 2 of this STAGE. Particular attention should be paid to airspeed and altitude control and remaining within the PTS limits. The following should be practiced:

____Steep turns.

____Slow flight. Straight and level and turns to headings, ±10°, ±100 feet, +10/-5 knots.

____Approach-to-landing stalls; straight ahead and turning.

____Takeoff and departure stalls. Straight ahead and turning.

____Descent for turns around a point.

____Turns around a point.

____Takeoffs and landings. Normal, crosswind, short- and soft-field, as applicable.

____Postlanding, shutdown, and airplane securing procedures.

____Review and discussion with instructor.

Assigned Reading. Student: *SPFM*, Chapters 19-25; PTS. Instructor: *FIM* Chapters 13 and 14.

Comments_____

_____

Instructor_____

Date_____Ground Instruction Time_____Flight Time_____

Student Initials_____

UNIT 4 - NAVIGATION

■ Ground Instruction. 1.0 Hour

____Pilotage and dead reckoning. Student plans a short cross-country calculating groundspeed, times and fuel required.

____Review of navigation systems and radar services.

____VOR bearings and cross-bearings and/or ADF homing to a station.

    ____Proper communications procedures when using ATC radar services.

____Diversion. Discuss diversions to alternate airports for the planned trip.

____Procedures when lost.

    ____Confess; climb; communicate; comply; conserve; cool, keeping of.

    ____Use all ATC communications available.

    ____Dragging the area.

■ Flight: Dual. 1.5 Hour.   Particular attention should be paid to the student's understanding of the following items:

____Preflight. Know and be able to explain *why* the various components of the airplane are checked.

____Use and understanding of checklist for prestart, start, and poststart procedures.

____External power source use and hot and cold weather starting.

____Cockpit management.

____Taxi, takeoff, and departure.

____Staying within altitude and heading limits during the cruise portion (*SPFM*, Chapter 28, PTS).

____VOR or ADF operations.

____Brief demonstration of procedures when lost.

____Awareness of geographical position.

____Communications with ATC or Unicoms.

____Return to airport.

    ____Checklist.

    ____Traffic pattern.

    ____Airspeed control.

    ____Landing. Touchdown and directional control.

____Postlanding and shutdown procedures.

____Evaluation.

____Critique.

____Review.

Assigned Reading. Student: *SPFM*, Chapters 12, 14, 15, and 28; Practical Test Standards. Instructor: *FIM* Chapters 10, 12, and 24.

Comments_____

_____

Instructor_____

Date_____Ground Instruction Time_____Flight Time_____

Student Initials_____

UNIT 5 - High Work and Emergency Instrument Flying

■ **Ground Instruction. 1.0 Hour**

____Review Unit 3, as necessary.

____Maneuvering during slow flight. Brief review of the power curve (*SPFM*, Fig. 12-13).

____Power-off stalls.

____Power-on stalls.

____Spin awareness.

____Six Basic Instrument Maneuvers, hooded.

    ____Straight and level flight.

    ____Constant airspeed climbs.

    ____Constant airspeed descents.

    ____Turns to headings.

    ____Recovery from unusual flight attitudes.

    ____Radio communications, navigation systems/facilities, and radar services.

■ **Flight: Dual. 1.0 Hour**

The student should practice the following:

____Slow flight.

____Power-off (approach-to-landing) stalls, straight and turning.

____Power-on (takeoff and departure) stalls, straight and turning.

____Instrument maneuvers, hooded.

    ____Straight and level.

    ____Constant airspeed climbs.

    ____Constant airspeed descents.

    ____Turns to headings.

    ____Recoveries from unusual flight attitudes.

    ____Simulated procedures when lost, with VOR tracking or ADF homing.

____Simulate (for a few hundred feet) an emergency descent (maintain airspeed ±10 knots).

____Orientation, division of attention and proper planning.

____High-altitude emergency approach and landing. Break off at a safe altitude unless it is planned for a landing at the home airport. Use proper communications and collision avoidance procedures in that case.

    ____Selection of suitable emergency landing area within gliding distance.

____Use of emergency checklist.

____Best glide attitude. Configuration and airspeed (±10 knots).

____Judgment of wind direction.

____Determine the reason for a malfunction and correct if possible.

____Flight pattern to selected landing area, considering altitude, wind, terrain, and obstructions.

____Positive control of the airplane at all times.

____Postflight instruction.

____Evaluation.

____Critique.

____Review

Assigned Reading. Student: *SPFM*, Chapter 28, Practical Test Standards.

Comments_____

_____

Instructor_____

Date_____Ground Instruction Time_____Flight Time_____

Student Initials_____

UNIT 6 - Systems and Equipment Malfunctions, Emergency Equipment, Survival Gear, and Night Operations

■ **Ground Instruction. 2.0 Hours.**   Review these subject items as indicated in *SPFM*, Chapter 28, and PTS. Make sure that the following items have been covered and checked off:

____Partial or complete power loss.

____Engine roughness or overheating.

____Carburetor or induction icing.

____Loss of oil pressure.

____Fuel starvation.

____Electrical system malfunction

____Flight instruments malfunction.

____Landing gear or flap malfunction.

____Inoperative trim.

____Inadvertent door or window opening.

____Structural icing.

____Smoke/fire/engine compartment fire.

____Emergency descent.

____Any other emergency appropriate to the airplane provided for flight test.

____Emergency equipment and survival gear.

    ____Location in airplane.

    ____Operation or use.

    ____Servicing requirements.

    ____Method of safe storage.

    ____Equipment and survival gear appropriate for operation in various climates and topographical environments.

    ____Emergency checklist.

____Night flying.

    ____Physiology. Rods, cones, and other factors.

    ____Lighting. Airplane, airport, and obstruction.

    ____Personal equipment essential.

    ____Night orientation. Navigation and chart reading techniques.

    ____Safety precautions and emergencies peculiar to night flying.

____Review general CRM and ADM as required for the private pilot.

____Evaluation

____Critique

____Review

Comments_____

_____

Instructor_____

Date_____Ground Instruction Time_____

Student Initials_____

UNIT 7 - STAGE Check: Practical Test

This flight is the final stage check in preparation for the student's FAA Practical Test. The check pilot will cover as much of the oral and flight portions as possible in the time allowed. Particular attention should be given to those weak areas as shown by the preparatory dual flights in this STAGE. The check pilot may, after this flight, call for extra time and a recheck.

■ Check Flight: Ground. 1.0 Hour; Flight Time. 2.0 Hours

____Evaluation.

____Critique.

____Review.

    ____Headwork.

    ____Air discipline.

    ____Attitude toward flying.

____Recommendation for the Practical Test.

____Recommendation for extra time and recheck.

Comments_____

_____

_____

_____

Endorsement for Flight Proficiency: FAR 61.107

I certify that I have given Mr./Ms._____ the flight instruction required by FAR 61.107 and find him/her competent to perform each pilot operation safely as a private pilot.

Date_____

Instructor Signature _____

CFI No._____ Exp._____

Check Pilot_____Cert. No._____Exp. Date_____

Date_____Ground Time_____Flight Time_____

UNIT 8 - FAA Practical Test

■ **Checklist.** Allow 2.0 hours ground and 2.0 hours flight for the Practical Test.

____Student pilot certificate current, plus medical certificate, if required.

    ____Student certificate properly endorsed for solo and cross-country.

____Logbook. All endorsements correct, including presolo written, solo, solo cross-country, stage checks and recommendations for Knowledge and Practical Tests.

____Knowledge Test results.

____Charts, *Airport/Facility Directory*, plus computers.

____Airplane papers and logbooks in order.

____Examiner's fee.

____Graduation certificate, if applicable.

____Application for practical test correctly filled out and signed.

Date_____

PASSED_____ Temporary certificate issued_____

FAA Inspector or Examiner _____

FAILED _____ Recheck required.

NOTES AND COMMENTS_____

_____

_____

_____

Student Initials_____